Springer Proceedings in Energy

The series Springer Proceedings in Energy covers a broad range of multidisciplinary subjects in those research fields closely related to present and future forms of energy as a resource for human societies. Typically based on material presented at conferences, workshops and similar scientific meetings, volumes published in this series will constitute comprehensive state-of-the-art references on energy-related science and technology studies. The subjects of these conferences will fall typically within these broad categories:

- Energy Efficiency
- Fossil Fuels
- Nuclear Energy
- Policy, Economics, Management & Transport
- Renewable and Green Energy
- Systems, Storage and Harvesting
- Materials for Energy

eBook Volumes in the Springer Proceedings in Energy will be available online in the world's most extensive eBook collection, as part of the Springer Energy eBook Collection. To submit a proposal or for further inquiries, please contact the Springer Editor in your region:

Kamiya Khatter (India)
Email: kamiya.khatter@springer.com
Loyola D'Silva (All other countries)
Email: loyola.dsilva@springer.com

David S-K. Ting · Ahmad Vasel-Be-Hagh
Editors

Engineering to Adapt

Proceedings of Engineering to Adapt 2023 Symposium and Industry Summit

Springer

Editors
David S-K. Ting
Department of Mechanical, Automotive and Materials Engineering
University of Windsor
Windsor, ON, Canada

Ahmad Vasel-Be-Hagh
Department of Mechanical Engineering
Tennessee Technology University
Cookeville, TX, USA

ISSN 2352-2534 ISSN 2352-2542 (electronic)
Springer Proceedings in Energy
ISBN 978-3-031-47236-7 ISBN 978-3-031-47237-4 (eBook)
https://doi.org/10.1007/978-3-031-47237-4

This Springer imprint is published by the registered company Springer Nature Switzerland AG
The registered company address is: Gewerbestrasse 11, 6330 Cham, Switzerland

Paper in this product is recyclable.

To everyone who aspires to adapt.

Preface

Byron Pulsifer declared "To adapt is to move ahead." The companion saying is asserted by Paul Lutus, "And the most successful people are those who accept, and adapt to constant change. This adaptability requires a degree of flexibility and humility most people can't manage." **Engineering to Adapt** is about eta, η, the seventh letter of the Greek alphabet, which is recognized as the symbol denoting efficiency or effectiveness. **G. T. Reader** literally defines **Engineering to Adapt** as **Waste Not, Want Not** in the opening chapter, challenging every one of us to do our part. To effectively realize Waste Not, Want Not, the issue of waste requires some clarity at the very least, as treating it with passivity may lead to a problem as serious as anthropogenic climate change. We cannot talk about waste reduction without considering waste generated from hotels. **O. B. Ezeudu** and **U. C. Ugochukwu** present the Nigerian perspective in Chap. 2, Characterization and Management Practices of Solid Waste Generated from Hotels in Awka City, Nigeria. They conducted interviews, field observation, and characterization analysis and quantification of solid waste in Awka City. It was found that among the studied hotels, the waste contained more than 44% organic content and over 94% recyclables. As such, simple policy changes, including the implementation of a circular economy, can be highly efficient in reducing waste. It is imperative to include combustion engine efficiency, along with clean and renewable fuel, when engineering to adapt. **B. Cong**, **S. LeBlanc**, **X. Yu**, and **M. Zheng** enlighten us with the latest update on this in Chap. 3, "Ignition and Combustion Characteristics of High-Pressure DME Spray Under Diluted Conditions." Their single-cylinder engine study sheds light on utilizing DME to realize ultra-low engine-out NOx and particulate matter emissions. A comprehensive comparison of six commercially available rechargeable batteries is furnished by **F. Balo** and **L. S. Sua** in Chap. 4, "A Critical Appraisal of Batteries with Metal Phosphate Among Commercial Batteries." They applied ten criteria to assess the six studied batteries with metal phosphate to deduce the most effective battery. The analysis also helped in identifying limitations and areas for improvement. Food is required to fuel the Engineering to Adapt. Employing a controlled environment in an agricultural greenhouse plays a critical role in sustaining this. **Q. Digweed** and **S. Sawan** expound on the "Impact of Phase Change Material on Greenhouse Energy

Balance Under Light Abatement Curtains," in Chap. 5. It is interesting to learn from their experiment using two identical 50 m^2 greenhouses that phase change material is only efficacious in saving heating during the cold season with the 16-hour lighting strategies and not with a 24-hour one. Regarding lighting, City slickers miss out on many a starry night because of overwhelming city lights. Trying to take away the beautiful night sky from country dwellers is a different story. **W. Lubitz**, **H. Henry**, **A. Nauta**, **S. Tasnim**, and **T. Graham** illuminate us with Chap. 6, "Comparative Assessment of Winter Night Sky Brightness in Southwestern Ontario." They measured sky brightness between fall 2022 and spring 2023 along a longitude of 42°N for an urban setting, a small town, a rural backdrop, and a rural area with nearby greenhouses. Their results suggest that supplemental greenhouse lighting appears to affect sky brightness but only in the immediate area. Work is being planned to differentiate the impact of anthropogenic from natural light sources on sky brightness. Renewable energy is an integral part of Engineering to Adapt. For effective greening of any energy usage, the option must be financially viable. **R. Babaei**, **D. S-K. Ting**, and **R. Carriveau** examine the feasibility of invoking solar tracking technologies (vertical axis, horizontal axis, and dual axis trackers operating with monthly and continuous adjustments) to further photovoltaic efficiency in Chap. 7, "Feasibility Analysis of Solar Tracking Technologies Connected to Renewable Energy Systems." For the studied rural residential setting in South Africa, the net present cost and levelized cost of energy of photovoltaic can be lowered by implementing a proper solar tracking and controlling strategy. With climate change, the need for refrigeration is going to increase. To sustainably cool, the energy-demanding compressor may be replaced, and this can be realized via a reversible adsorption process. Metal-organic frameworks are potential adsorbents that can significantly outperform rivalries such as silica gel and activated carbon. In Chap. 8, "Metal Organic Frameworks (MOFs) in Adsorption Heat Transformations," **J. Al-rabadi** and **J. Aman** present a state-of-the-art review on metal organic frameworks in heat adsorption transformation, including their superior performance and outstanding challenges. Cooling is not just a necessity for human thermal survival and thermal comfort, the hardworking wind turbine electromechanical equipment enclosed inside the nacelle also needs to be kept cool. **B. Sen** and **N. Datta** disclose a three-dimensional CFD simulation of the airflow pattern and temperature variations inside the nacelle of a small horizontal axis wind turbine in Chap. 9, "Thermal Analysis of the Nacelle of a Small Horizontal Axis Wind Turbine Using a CFD Model in ANSYS-FLUENT." They reveal that natural cooling by directing outside air into the nacelle can keep it sufficiently cool. Engineering to Adapt must include measures to keep our environment clean. **S. Narimannejad**, **Q. Cai**, **B. Zhang**, **K. E. Taylor** and **N. Biswas** target the harmful cobalt in Chap. 10, "Parameter Interactions on the Adsorption Behaviour of Cobalt onto Saline Soil with Different Biosurfactants." The adsorption behavior of cobalt onto saline oil with various biosurfactants was studied, and the conditions for the highest cobalt washing efficiency were deduced. This volume concludes with "Real-Time Optimization of Yaw Angle and Tip-Speed Ratio for a Six-Turbine Plant of NREL 5-MW Wind Turbine," by **A. Hosseini**, **D. T. Cannon** and **A. Vasel-Be-Hagh** as Chap. 11. Back to engineering efficiency, this chapter aims at the recovery of energy loss due to upstream wind turbines in a wind farm. Their case study shows that simultaneous control of turbine yaw and tip-speed ratio can result in significantly larger efficiency gain compared to either control strategy operating in isolation.

TURBULENCE & ENERGY LABORATORY

CONFERENCE SERIES

Engineering to Adapt

2023

Windsor, Canada
Cookeville, USA
September 2023

David S-K. Ting
Ahmad Vasel-Be-Hagh

Acknowledgments

This book could not have materialized if not because of the faithful support of various conscientious stakeholders. The editors continue to enjoy working with the marvelous Springer Nature team led by Anthony Doyle. It is a privilege to participate in bettering tomorrow striving with like-minded colleagues. The quality contributions from the outstanding experts who fashioned the individual chapters. And a heartfelt Thank You goes to the reviewers who executed quality control of every chapter.

September 2023

David S-K. Ting
Turbulence and Energy Laboratory, University of Windsor
Ahmad Vasel-Be-Hagh
Fluid Mechanics Research Laboratory, Tennessee Tech University

About This Book

Eta, η, the 7th letter of the Greek alphabet, is scrupulously used to denote efficiency, which is what *Engineering To Adapt (ETA2023)* strives for. In context, efficiency, η, is about avoiding waste, may this be energy, time, money, or material, in accomplishing something useful. As such, *ETA2023* aims at bringing experts and emerging leaders together to innovate more efficient ways to engineer and live. In other words, *ETA2023* strives to synergize and catalyze all stakeholders, enthusiasts, and experts from academia, industry, policy arenas, and the general public, to formulate novel ways to improve tomorrow. This symposium will disseminate recent progress and promote collaborations to maximize opportunities for innovative integrated solutions. Topics of interest include resource and energy efficiency, waste reduction, and eco-friendly agriculture, architecture, engineering, and living.

Contents

neering to Adapt: Waste Not, Want Not 1
iam T. Reader

aracterization and Management Practices of Solid Waste
enerated from Hotels in Awka City, Nigeria 55
Obiora B. Ezeudu and Uzochukwu C. Ugochukwu

3 Ignition and Combustion Characteristics of High-Pressure
DME Spray Under Diluted Conditions 63
Binghao Cong, Simon LeBlanc, Xiao Yu, and Ming Zheng

4 A Critical Appraisal of Batteries with Metal Phosphate
Among Commercial Batteries 79
Figen Balo and Lutfu S. Sua

5 Impact of Phase Change Material on Greenhouse Energy
Balance Under Light Abatement Curtains 95
Quade Digweed and Sabrina Sawan

6 Comparative Assessment of Winter Night Sky Brightness
in Southwestern Ontario 113
William Lubitz, Heather Henry, Alex Nauta, Syeda Tasnim,
and Thomas Graham

7 Feasibility Analysis of Solar Tracking Technologies Connected
to Renewable Energy Systems 125
Reza Babaei, David S.-K. Ting, and Rupp Carriveau

8 Metal Organic Frameworks (MOFs) in Adsorption Heat
Transformations 157
Joud Al-rabadi and Julia Aman

9 Thermal Analysis of the Nacelle of a Small Horizontal Axis
Wind Turbine Using a CFD Model in ANSYS-FLUENT 173
Bhaswati Sen and Nabanita Datta

10 Parameter Interactions on the Adsorption Behaviour of Cobalt onto Saline Soil with Different Biosurfactants 191
S. Narimannejad, Q. Cai, B. Zhang, K. E. Taylor, and N. Bisw

11 Real-Time Optimization of Yaw Angle and Tip-Speed Ratio for a Six-Turbine Plant of NREL 5-MW Wind Turbine . . . . . .
Amir Hosseini, Daniel Trevor Cannon, and Ahmad Vasel-Be-Hagh

Chapter 1
Engineering to Adapt: Waste Not, Want Not

Graham T. Reader

Abstract In the not-too-distant past there was a sufficiency of natural resources and many fewer people to consume them. In the global north, this situation led to societal behaviour which can be described as *make-use-discard* and applied not only to the basic needs for human survival, i.e., drinking water and adequate food, but also to generated energy and a whole range of manufactured products. One of the outcomes of this mentality was the creation of enormous amounts of waste, which continues today. The abundance situation is now being reversed, caused largely by an eight-fold increase in global population compared with that existing over two centuries ago. This increase has been accompanied by diminishing reserves of natural materials and a growing need for waste disposal. Arguably, this situation makes the idiom *waste not want not* even more relevant today than it was when it first appeared in the sixteenth century. As more waste measurement data are becoming available and populations, especially in low-income counties, continue to rise, the premise that national income levels are the only driver of per capita waste is no longer unquestionable. Indeed, the modern rapid population growth in poorer countries will likely result in increasing quantities of global waste, largely unmanaged. Consequently, it has been suggested that, if global action is not taken soon, waste from future generations of human activities could easily overwhelm the Earth's natural environmental systems. Nevertheless, in comparison with the multitude of issues facing society (e.g., anthropogenic climate change, extreme poverty, disaster relief and reduction) is waste generation such a serious problem? If it is, what additional actions need to be taken? The topics discussed in this paper are the current waste related problems, what is being done to reduce them, and what could be done to prevent severe ecological damage in the future.

This chapter is partially based on materials used in a Keynote presentation delivered 22 June 2023 at the 'Reclaiming Eden – Engineering a Brighter Tomorrow Symposium' in Windsor, Ontario, Canada.

G. T. Reader (✉)
Department of Mechanical, Automotive and Materials Engineering, University of Windsor, 401 Sunset Avenue, Windsor, ON N9B 3P4, Canada
e-mail: greader@uwindsor.ca

D. S-K. Ting and A. Vasel-Be-Hagh (eds.), *Engineering to Adapt*, Springer Proceedings in Energy, https://doi.org/10.1007/978-3-031-47237-4_1

Keywords Waste · Waste Management · Climate Change · Adaptive Engineering

Abbreviations

BEV	Battery Electric Vehicle
CO_2	Carbon Dioxide
CH_4	Methane
CU	Consumptive coefficients (semi-empirical)
EfW	Energy from Waste
EU	European Union
FEWS	Famine Early Warning System
FLI	Food Loss Index
FLW	Food Loss and Waste
FWI	Food Waste Index
GDP	Gross Domestic Product
GHG	Greenhouse Gas
GNI	Gross National Income
GWP	Global Warming Potential
HEV	Hybrid Electric Vehicle
IPC	Integrated Food Security Phase Classification
LDC	Least Developed Country
LFG	Landfill Gas
LIFDC	Low Income Food Deficit Countries
MSW	Municipal Solid Waste
NDC	Nationally Determined Contributions
NGO	Non-government organization
PACE	Platform for Accelerating the Circular Economy
SDG	Sustainable Development Goal
SEEA	System of Environmental-Economic Accounting
UN	United Nations
UNEP	United Nations Environmental Program
UNFAO	The Food and Agriculture Organization of the United Nations
UNICEF	United Nations Children's Fund
USAID	United States Agency for International Development
USEPA	United States Environmental Protection Agency
USFDA	United States Food and Drug Administration
USHHS	United States Department of Health and Human Service
VCB	Vegetated Compost Blanket
WBG	World Bank Group
WEF	World Economic Forum
w-e-f	Water-energy-food
WFP	World Food Programme
WHO	World Health Organization

WTE	Waste-To-Energy
WUE	Water-use efficiency
ZWIA	Zero Waste International Alliance
UK	United Kingdom
US	United States

1.1 Introductory Remarks and Sources of Waste

On 17 October 1987, a large mass gathering in Paris, France assembled "*to honour the victims of extreme poverty, violence and hunger*" [1]. In 1992 this date was adopted by United Nations (UN) resolution as the 'International Day of the Eradication of Poverty' [2]. It is now its third decade (2018–2027) of the UN's formal implementation of resolutions for eliminating poverty [3]. What is the connection between the idiom *waste not, want not*, and poverty eradication? One helpful meaning of the idiom is that a 'wise use of resources will prevent poverty' [4]. It was this association that was emphasized directly or indirectly in the three interrelated landmark UN resolutions adopted in 2015, addressing anthropogenic climate change,[1] a framework for reducing or preventing disaster risks, and a universal transformation to sustainable development [5–7]. Since that time, it has become increasingly evident that despite the many measures being taken to fulfil these global commitments it is unlikely that they will be realised in their entirety by the 2030 target date especially with regard to reducing waste by 50% [7]. Nevertheless, in numerous instances some *progress* has been made [8–10]. A notable example of such progress is the reduction of the percentage of the global population suffering extreme poverty, especially in India and China, as shown in Fig. 1.1 [11]. However, care must be taken with such measures because the percentage reduction could be more than offset by population growth, such that, more people may live in extreme poverty, but that group, as a percentage, is lower than before.

In terms of climate change, especially the dreaded consequences of global warming, the underlying strategy has been mitigation involving the rapid transition to renewable and sustainable energy sources as complete replacements of traditional carbonaceous fossil fuels, namely coal, oil, and natural gas. Prior to the hyperfocus on combating climate change, the main initial arguments for seeking alternatives to fossil fuels were the purportedly diminishing proven reserves of such sources. This line of reasoning has gained far less traction than the attribution of fossil fuel greenhouse gas (GHG)[2] emissions, especially carbon dioxide (CO_2) and methane (CH_4), being the major cause of climate change as manifest by global warming. The GHG impacts on climate are largely unchallenged in the scientific community

[1] A description foreshortened to climate change in the remainder of the chapter.

[2] A table of abbreviations and their meaning are given in Table A before the chapter references.

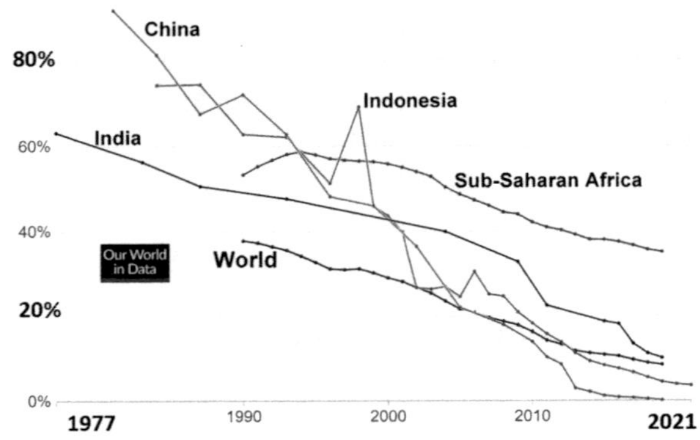

Fig. 1.1 Exemplars of extreme poverty reduction as percentage of population [11]

but the magnitude of their influence and the routine dismissal of other likely additional causes of global warming remain contentious issues, albeit rarely among many political and media communities.

Not surprisingly then other issues that some associate with climate change have not attracted the same degree of attention or the same scale of research funding as the detrimental climate effects related to the continued use of fossil energy sources. This, arguably inflexible approach, is now changing as, for example, food waste has been identified as a major generator of greenhouse gases (GHG) and therefore a key factor in the creation of climate change [12, 13]. Nonetheless, this recent focus of the connection between food waste and climate change does not mean waste generation and handling has been a neglected topic, as evidenced by the significant body of literature dealing with such issues e.g., [14–16]. The need to reduce waste in connection with climate change will be discussed in more detail in Sect. 1.4 of this chapter.

Historically, severe weather events, e.g., hurricanes,[3] forest fires, droughts, and floods, have habitually led to infrastructure damage and loss of life and continue to do so. From written records and oral accounts, originating long before the age of fossil fuels, disastrous natural occurrences involving earthquakes, impacts with astronomical objects, volcanic eruptions, tsunamis, and the vagaries of orbital cycles have impacted both short-term weather patterns as well as longer lasting natural climate changes [17]. These events are often grouped under the collective terms of ‘climate variabilities’. However, today many of the short-term severe weather episodes are considered by authoritative sources to be increasing in frequency and scale due to climate change, e.g., [18]. To combat the devastating impacts of both natural and human-enhanced weather events, the Sendai disaster reduction framework resolution was passed in 2015 by the UN General Assembly [6]. This framework has 7 global targets and 38 indicators mainly aimed at better forecasting the risk of pending disasters and, when natural disasters[4] inevitably occur, for provisions to be taken

[3] Also known as typhoons and tropical cyclones [https://oceanservice.noaa.gov/facts/cyclone.html].

[4] The United Nations Office for Disaster Risk Reduction (NDRR) prefer the term natural hazard.

for '*building-back-better*' infrastructure. The aim being to significantly reduce the scale of damage from any future disaster occurrences, e.g., building higher sea-wall defences [19, 20]. Obviously, if reducing certain types of waste forestalls, or weakens, damaging weather events, there could be measurable economic, as well as societal, advantages connected to the lessening of population displacements and infrastructure deterioration. Nonetheless, even in the absence of climate change, at times there will inevitably be damaging natural catastrophes and weather disasters resulting in all types of waste being produced, from degraded buildings, transportation, energy, and service infrastructures, to crop failures. Should the waste from natural disasters be recycled, repurposed, or collected and dumped into landfill sites?

Of the three landmark UN agreements, the sustainability transformation is aimed specifically at the total elimination of global poverty. To achieve this worthy ambition will require the attainment of 169 'targets' associated with 17 defined sustainable development goals (SDGs) [7]. It could be anticipated that, of this large number of sustainability targets, many would be associated with waste reduction, but only 7 embody the term 'waste' or similar words such as wastewater. For example, one of the waste targets, SDG 12.c, is to encourage global reductions in the wasteful consumption of fossil-fuels by rationalizing "*inefficient fossil-fuel subsidies*" [7]. It is not clear what the difference is between inefficient and efficient subsidies. Arguably, it could be implied that efficient subsidies do not cause wasteful consumption of the particular fuel. However, in a similar manner to the ambiguities encountered with the definitions and uses of the word, 'subsidy', application of the term 'waste' can also be a problematic characterisation. Indeed, the word *waste* has a multitude of meanings as have many other words in the English language, e.g., *bank,* where the actual meaning of the word relies on the context. Lexicographers call such words, polysemous [21].

Phrases such as *a waste of time* or *a waste of space*, can be used in a disparaging sense when describing an individual, painting a different picture from the use of the word waste when dealing with food, water, energy, and other natural resources. The etymology of the word 'waste' dates back to as least the twelfth century CE, since when the original meanings of '*to squander, spoil and ruin*' have evolved to also infer; to spend or consume uselessly, damage, destroy, kill, desolate land, wilderness, and, in slang terms, to describe a state of intoxication from alcohol and drugs. It is important then, at least for the sake of this chapter, to identify the contexts in which the term *waste* is used. However, simply describing the circumstances alone may not be sufficient for a complete appreciation of the use of the word. For example, if an individual, for whatever reasons, does not wholly consume their meal, but the *left-overs* provide a meal for someone else, does this constitute an example of food waste? It may do, especially if not all the food is eventually consumed, but in which case how can the amount of the food waste be measured and by whom? These are not straightforward tasks, indeed, in general, precisely defining and accurately measuring waste generation and disposal are challenging matters. Thus, the discussion of waste can be further complicated by considering the implications of the idiom '*one man's junk is another man's treasure*'. This expression, if revised, by replacing the word *junk* with *waste* and *man* by *person*, would indicate that, as with the food example, that certain types of waste can be reused, recycled, or repurposed for further consumption by others, or indeed, by the original waste generator. In harmony with the idiom, other items which could be reused include discarded clothes which no longer fit, or which to some are no longer fashionable, but that can be worn by others [22]. Moreover, *old* clothes or fabric scraps can also be recycled and repurposed to create floor and wall coverings, i.e., rag-rugs, and workshop cleaning rags or household dusters.

Other examples of the utility of notionally discarded waste are the use of old wooden flooring, taken from a house to be demolished, in the construction of a new or renovated home and the reuse, recycling or repurposing of other building materials, decorative or structural, such as bricks and stones, in similar circumstances [23]. There are likely numerous other repurposing examples but perhaps one of the more obvious approaches to reducing waste is to repair items which would otherwise be discarded. Interestingly, some items which are still functional, e.g., smartphones, are neither discarded or repurposed when replaced by newer improved versions nor passed-on to other users [24]. If any discarded items are reclaimed, or even hoarded, is the use of the word waste appropriate in such cases? Is then the measurement of the amount of waste that is sent to a 'dump'[5] e.g., a landfill site, indicative of the total amount of waste being generated? Once again, the answer is by no means simple as the waste that is dispatched to a landfill site is usually called Municipal Solid Waste (MSW) and does not include the natural biological waste produced by humans in the form of urine and faeces, sometimes referred to as 'human waste' [25]. MSW accounts for about 2 billion tonnes of waste globally whereas human faeces, before any purposeful mixing with water, amounts to the equivalent of about 10% of MSW [26]. Human produced natural biological wastes are then collected, in many countries, directly from households, commercial and industrial buildings in sewage piping and drainages systems after being mixed with water, which is often unnecessarily high-quality drinking water. This 'wastewater' may be then treated to produce water deemed safe for reuse.

In instances where piped sewer systems are not available there may still be other means of hygienically separating excreta from human contact. When some form of acceptable separation is available the term used to describe such systems is 'improved sanitation facilities', if such a facility is not shared with other households and the excreta is disposed of in-situ or transported for treatment off-site, as described in the previous paragraph, the facility is described as 'safely managed' [27]. The problem is that 3.6 billion people globally lack access to safely managed sanitation services and between 6 and 8% of the world's population still practice open defecation in field, forests, beaches and other open spaces [27, 28]. While the world, in general, has made progress in providing various levels of sanitation facilities to more people, that success is not reflected in the sub-Saharan region, in particular in Eastern and Southern Africa, where an increasing number of people do not have access to even the most basic sanitation services [29]. The technology exists to address these issues, but economic and political obstacles preclude its wholesale implementation. Maybe

[5] In the western vernacular landfill sites are often to referred to as dumps although the term is used differently in global analyses.

innovative engineering designs which are less expensive and could be operated and maintained by local workforces may persuade more political investment in the sanitation problem, but to attract more foreign, and particularly, World Bank financial support the designs have to use renewable energy sources [30].

1.1.1 Types of Waste

MSW and sewage sludge are only two forms, or categories, of waste. If waste is to be reduced, then all categories of waste generation must be included. But what are they? Many national and international, governmental and non-governmental, agencies and organizations have encyclopedic lists and definitions of the types of waste generated by humans and as a result of human activities, but a basic framework is provided by the United States Environmental Agency (USEPA) as shown in Table 1.1 [31].

An idea of the breadth of the various categories can be gauged by the fact that MSW is a collective term for at least 13 different waste streams including food, metals, textiles and so on, Table 1.2 [32]. Another example is the 180-page document issued in the United Kingdom dealing with the classification and assessment of numerous types of hazardous waste [33]. These are then but two examples for each type of waste category that have a multitude of different classifications and characteristics. Subsequently, how to reduce all waste is a topic of immense scale and the challenges of achieving zero-waste is comprehensively acknowledged by many national and international organizations, as evidenced by the announcement of the UN General Assembly that, from 2023, the 30th March will be known as the "*International Day of Zero-Waste*" [34]. It remains to be seen whether this call for international action to reduce global waste by 50% by 2030 and 100% by 2050 will have the desired effect, for the targets seem to be unduly optimistic, or are they? Why is waste generated in the first place?

Arguably, the generation of most waste products is not malevolently intentional, but is largely the result of the 'take-make-use-discard' way of life (Fig. 1.2) which, in modern times, has flourished, especially in this age of consumerism [35, 36]. However, there are situations, food and medicines being obvious examples, where

Table 1.1 USEPA waste categories [31]

USEPA waste categories	
MSW	Radioactive
Hazardous	Construction and demolition debris
Industrial non-hazardous	Extraction and mining
Agricultural and animal	Oil and gas production
Medical	Fossil fuel combustion
Sewage Sludge	

Table 1.2 MSW materials generating streams (United States) [32]

MSW—materials generated	
Paper and paperboard	Rubber and leather
Glass	Textiles
Metals:	Wood
1. Ferrous	Food
2. Aluminum Metal	Yard trimmings
3. Other Non-Ferrous Metal	Miscellaneous inorganic wastes
Plastics	Other (includes battery electrolytes, fluff[6] pulp, feces and urine in disposable diapers but does not include construction and demolition debris, industrial process waste and certain other wastes)

because of the inability to safely store the products, i.e., lack of access to refrigeration storage facilities, they have to be discarded. In other instances, especially with unwanted, unused, or out-of-date pharmaceuticals, depending on the healthcare environment and infrastructure in a particular country or region, the medicines are dumped in the trash or a convenient water source [37]. Even in a rich country like the United States it may not be possible to dispose of unwanted medicines other than by flushing them down the toilet or putting them into household trash, but the U.S. Food and Drug Administration (USFDA) have published 'flushing' and 'non-flushing' lists [38]. It is not just unwanted pharmaceuticals that find their way into wastewater systems because human excreta may also contain potentially harmful substances from prescription drugs taken as part of a medical treatment regimen [39]. Whatever the source of these particular waste products they are giving cause for concern because of the potential risks they present to human and aquatic life as well as ecosystems [40]. The increased global use of prescription pharmaceuticals is creating serious challenges to existing wastewater treatment systems, even the most modern and sophisticated versions, and there is an urgent need for more scientific investigations coupled with the development of new technologies if these challenges are to be overcome [41]. Even if wastewater treatment systems can combat such issues successfully it seems unlikely that human healthcare organizations will restrict or reduce the future use of prescription drugs, but it could be possible that scientists and engineers may find a way of recycling extracted pharmaceuticals.

If the take-make-use-discard approach itself is to be totally dismantled then, for engineers in particular long familiar with the industrial philosophy of planned obsolescence and the commercial and medical employment of single-use items, such a change would involve a firmer adoption of the 'cradle-to-cradle' approach in the

[6] Fluff pulp is chemical pulp made from long-fibre softwoods.

Fig. 1.2 Schematic of take-make-use-discard lifestyle

design, manufacture and operation of equipment [42]. Fortunately, in many engineering industries, especially in the land-based automotive sector, building equipment whose component parts can be reused, recycled, or repurposed is hardly a new experience [43]. Indeed, there is now legislation in several countries which requires vehicular components to be inherently recyclable once a vehicle has reached the end of its useful life [44]. However, the increased use of plastics in vehicular construction in recent years may strain a manufacturer-assembler's ability to comply with the legal regulations as will the need to recycle the lithium batteries used in hybrid and battery electric vehicles (HEV and BEV) [45, 46].

1.1.2 MSW: Energy, Composting, Recycling Materials

Regardless of the type of waste that is generated by human activities, the provision of appropriate infrastructure plays a crucial role in the ability to handle and manage waste. This infrastructure, where it exists, will include facilities for collecting and processing waste of all kinds. For example, after it is collected not all MSW will automatically be sent to a regulated landfill site, of which there are several types (as briefly discussed in Sect. 1.5). If the waste is non-recyclable, it may undergo one of several process to produce energy in the form of heat, electricity, or fuel. Additionally, if the MSW is deposited in a landfill site, usually purposely designed, gas, mainly methane, can be generated by anaerobic digestion and combusted in-situ or at an offsite location to produce energy. All such processes are frequently referred to as 'waste-to-energy', WTE or EfW [47, 48]. There is a lengthy history of burying or burning MSW and waste incineration was not always aimed at producing energy, but often done to save land space in heavily populated areas and regions, such as Japan [49, 50]. Whatever the reasons for burning waste, since the 1960s, with the advent of societal concerns about air quality, the associated processes have faced increased scrutiny [47, 51, 52].

Nevertheless, China continues to increase the amount of waste that is burned and has the largest EfW capacity globally which, over a 5-year period to 2016, grew at an average rate of over 26% to 7.3 GW [48]. This level of capacity is almost equivalent to all the solar and wind installed power capacity of Canada's most populous province, Ontario [53]. However, in the Chinese context the EfW capacity represents less than one-third of a percent of their installed energy capacity [54]. In this case, burning or incinerating waste is more of a method of reducing the volume of waste rather than a concerted effort to provide an addition to the country's energy mix. Conversely, Sweden uses waste-to-energy to produce heat and electricity for a large number of households, maybe as many as 40%, while their government acknowledges that CO_2 is generated, the GHG emissions from landfill sites are considered to be of relatively greater climate change impact, albeit thinking which is questioned by European Union (EU) officials [55–57]. Notwithstanding such concerns, there are at least 47 countries globally employing WTE facilities with over 50% of the almost 2200 installations located in Japan, but only 1 in the whole of sub-Saharan Africa [50].

For a variety of reasons, it is not possible to achieve 100% efficiency in any combustion process, so there are always some unburned by-products, in the case of waste incineration 'ash' is generated. The bulk of this ash is called 'bottom ash' (Fig. 1.3) from which any metallic compounds can be separated, the remaining ash may contain materials suitable for use as aggregate in road and rail construction or be mixed with 'fly ash' and deposited in a landfill site [47, 58]. Fly-ash is fine particulate matter which either gets trapped using electrostatic precipitators or is filtered using so-called bag-filters, or both, from the flue-gases, as indicated in Fig. 1.3 and is wet-sprayed prior to mixing with the bottom ash. The actual constituents of fly ash and bottom ash depends upon the original composition of the MSW feedstock, and several studies have found that the heavy metal content of fly ash is greater than in bottom ash, so further treatment options need to be implemented prior to landfill disposal [59].

The term fly ash has long been associated as a residue from coal burning thermal power plants and the systems for extracting the fly ash are almost identical to that shown on Fig. 1.3 for MSW incineration [60]. Coal-burning derived fly ash is used in several engineering applications as a filler for highway asphalt, especially Portland cement concrete [61]. However, as there are global efforts by many countries to reduce and eventually eliminate coal use as part of their climate change mitigation strategies, could incinerated MSW fly ash be used as a replacement? In this millennium this option is being increasingly explored [62–64]. The main point is that, the amount of MSW sent to landfills can be reduced because portions of that waste can be recycled, reused, or repurposed; the key 'R's of sustainability. Nevertheless, while such waste management systems can alleviate the negative environmental impacts of MSW they do not significantly reduce the amount of MSW generation. Similarly, composting organic waste materials in MSW is more of a handling and management tactic rather than a way of reducing the amount of such generated waste. Even so, as particular constituents of MSW are separated for recycling and reuse their net effect is to reduce other forms of waste generation.

For instance, if energy is produced from MSW then the amount of energy supplied from non-renewable sources can be reduced but determining the scale of these 'savings' on a global basis remains elusive, although the Swedish energy example provides an insight into what can be achieved. Another example is that the retrieval

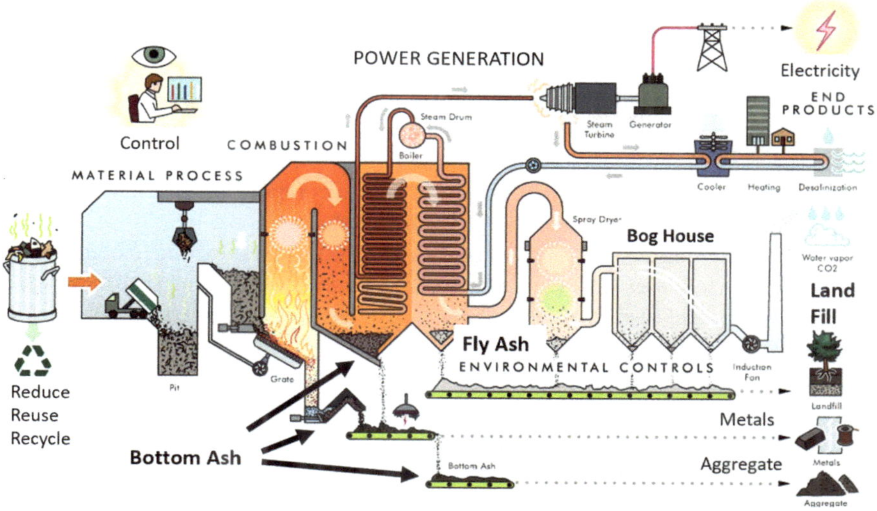

Fig. 1.3 Waste-to-energy system—deltaway, author modified [58]

and recycling of metallic elements could reduce the amount that has to be extracted by mining operations. This type of opportunity is likely to become increasingly important since one of the emerging growing waste problems globally is electrical and electronic wastes (e-waste), such as cell[7] phones, televisions, laptops, tablets, watches. Only 35% of e-waste is recycled with the final destination of the remainder, 65%, being a landfill site, yet according to a World Economic Forum (WEF)—Platform for Accelerating the Circular Economy (PACE) report, "*[t]here is 100 times more gold in a tonne of mobile phones than in a tonne of gold ore*" [65]. Unfortunately, as previously mentioned, there is a reluctance to discard unwanted electronics [24, 66].

One way of improving the nature of soil, especially for the growing of vegetation, food, trees, and fuel-crops, is the introduction of composted organic materials into the soil structure. In essence, the controlled decomposition of organic material, i.e., composting, provides a natural fertilizer. There are two main types of organic waste materials which are used in compositing, so-called 'Green' and 'Brown' [67]. Green organics are food waste, certain yard and garden trimmings, i.e., grass and plant clippings, and manure, whereas Brown organics contain other yard trimmings such as dry leaves, twigs, and wood chips, plus paper and shredded cardboard [67, 68]. As can be seen from Tables 1.1 and 1.2, MSW streams involve all these organics including small amounts of human manure from disposable diapers. The green portions have high levels of nitrogen whereas the brown constituents contain similar amounts of carbon, a balance between the two is said to provide "*effective composting*" [68]. Some local authorities and municipalities ask households to separate their collected

[7] Referred to as mobile phones or smartphones in some countries.

waste into different colored boxes for paper products and plastics and metals, and into large paper bags for yard and garden trimmings but few, as yet, are able to handle food waste other than in normal trash bags or containers.

As food waste is not only a significant proportion of MSW generation, but also a beneficial composting material, some agencies, such as governmental, professional societies, commercial associations, environmental and particular interest groups, urge households to use food waste as part of in-situ composting in lieu of dedicated collections and recommend ways to generate suitable compost [69, 70]. Gardeners and some smaller agricultural enterprises have long used compositing to help them in their fertilizing activities. There are some off-site larger scale commercial composting facilities which can also use parts of the waste streams identified in Table 1.1 to manufacture specific composts and, as with domestic composting, 'how-to' guides have been available for some time [71, 72]. Apart from the production of natural fertilizer the use of compost in 'vegetated compost blankets' (VCBs) has, and is, being investigated in the United States for service as stabilizers of highway embankments and stormwater control structures [73]. This is an example of the innovative and entrepreneurial effects to reduce waste that largely go unnoticed in the media or outside specialized technical communities. However, such ignorance, or lack of acknowledgement, also extends to other sectors such as those living in residential dwellings. The many promptings of authorities encouraging for this particular sector to recycle more food waste seems to be largely ignored especially in a high-income country like the United States. In 2018, only 3% of food waste from households was used for composting, maybe due to lack of separated collection and processing facilities, while 81% of wasted food was either landfilled or sent to a sewer, sufficient to feed between 75 and 92 million people for a whole year! [74, 75] This amount would enable Latin America, the Caribbean and Pakistan to eliminate hunger and undernourishment from their impacted populations [76].

In the, member approved, UN plan to achieve global sustainable development, one of the three key interlinked 'nexus' elements are water-energy-food (w-e-f) [77, 78]. The global food system is also embedded in other 'nexus' initiatives such as health-biodiversity-climate action [79]. There are other 3- and 4-element nexuses associated with sustainable human development, but in this chapter, it is the w-e-f nexus that is of interest because of its obvious connection with waste reduction. It is not an exaggeration to state that the efficient use of w-e-f resources is vital for human survival. Without question, apart from breathable air, humans can only live for a few days without water, maybe a month without food and, depending upon prevailing weather, climate changes and living conditions, an uncertain amount of time for those almost wholly dependent on access to electricity and the numerous basic services powered by electricity [80]. Fossil fuels, mainly natural gas and coal, provides the energy sources for about 2/3 of the world's electrical power and, in the United States, about 80% of its annual energy consumption is supplied by the combustion of traditional fossil fuel sources [81]. Of course, for the envisaged sustainable future, the intention is that fossil fuels will be largely replaced by renewable energy sources, but the point to be made here is that when all types of energy are converted to usable power, the bulk of the converted energy will be wasted, in the case of the United

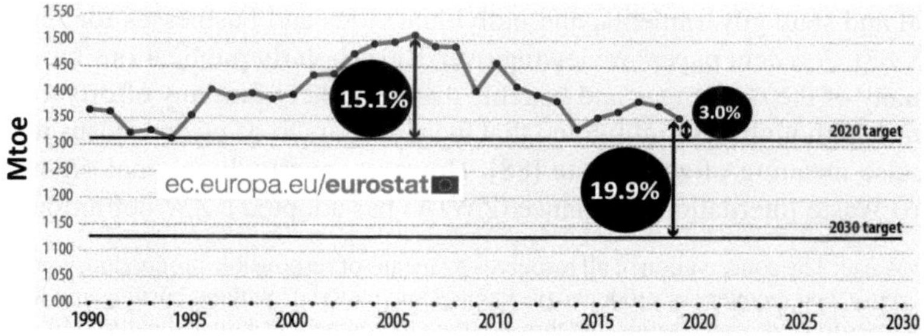

Fig. 1.4 European union energy consumption targets and actual usage [84]

States this is about 66% [81]. Much of this wasted energy is rejected due to the inherent efficiencies of the conversion processes which are governed, largely, by the laws of thermodynamics and physics and the limitations of available materials.

Nevertheless, conversion efficiencies can be improved and the wasted or rejected energy can be recuperated or reused for other purposes but, clearly, there are practical and scientific limitations. Many jurisdictions have stated targets for improvements in energy use efficiency and some, such as the EU and the European Union Commission, have mandated the amounts by which energy efficiency must be improved compared with a 2007 baseline, e.g., 20% by 2020 and 32.5% by 2030, using primary energy consumption as the measure, albeit that the 2030 target has been recently increased to 42.5% [82, 83]. Political edicts and engineering realities are not always in harmony but to date the results appear to have been fairly impressive if not perfect, although the policy-makers are not satisfied with the progress being made (Fig. 1.4) [84]. Arguably, not only is reducing energy waste by improving energy efficiency a sound climate mitigation tool but it should also help future populations in growth regions meet their energy needs. But, as always, waste reduction is impacted by other factors than political and technical, such as societal acceptances and awareness as well as economic realities [85].

1.1.3 Zero-Waste

The discussion of waste generation and utilization has so far focused on MSW as this form of waste involves all humans, whereas only some will be involved in the other categories of waste as given in Table 1.1, apart from sewage. The methods outlined for the handling and treatment of MSW, and the technologies involved, are particularly associated with the countries who have the economic capabilities to implement them, but for many low-income countries and regions they remain inaccessible. This is particularly worrisome for the future as many such countries are forecast to have far higher rates of population increases and, in some cases, almost triple their current amount of waste generation by 2050, the acclaimed target year by many NGOs, national and state governments, and individual cities and businesses for zero-waste [16, 86, 87]. A recent paper[8] reviewing over 100 scholarly publications has summarized many of the challenges and benefits that will likely face any efforts to achieve zero-waste although it is highlighted that there appears to be no consensus regarding the precise meaning of zero-waste [88]. However, one multinational organization, the Zero Waste International Alliance (ZWIA) has adopted a ZW definition,

> Zero Waste: The conservation of all resources by means of responsible production, consumption, reuse, and recovery of products, packaging, and materials without burning and with no discharges to land, water, or air that threaten the environment or human health. [89]

In essence, this succinct statement captures the characteristics of SDG 12 and its targets but the 'all' and 'no' objectives, while laudable, are perhaps far more ambitious than the UN's transformation agenda, at least in the latter's timeframe, i.e., 2030 [7].

While MSW is a universal waste issue, it is also apparent that it can be viewed as a resource since useful by-products such as energy, metals, structural fillers, and compost could be generated from it using appropriate technologies. These possibilities and opportunities require the necessary infrastructure to facilitate a 'waste management supply chain' from collection to treatment to utilization. Higher-income countries have the wherewithal to provide such infrastructure but what about poorer countries and regions? Is then the generation of waste and its reduction only an issue for those countries whose socio-economic philosophies are based on 'take-make-use-discard' and 'planned obsolescence' approaches? Does population size play a role in the amount of waste generation? Instinctively, it could be presumed that rich countries will generate more waste per capita than those with lower wealth levels and, for countries at the same income level, those with larger populations will generate waste more than those with fewer people, but how accurate are these inferences? These type of queries and their underlying assumptions are discussed in Sect. 1.2. Despite the highlighting of MSW it would be remiss not to give some attention in the remaining paragraphs of this section on how the 'waste not: want not' concept also applies to the provision of sufficient amounts of safe air and access to safe water, both needed for human survival. But air and water alone cannot guarantee survival because an adequate amount of food is also a prime requirement. However, there is a difference between food production and consumption involving another component of the global waste mixture and that is food waste. Food is not only a survival necessity but is intimately linked with MSW. Thus, the connections between the two are discussed in Sects. 1.3 and 1.4, along with a general overview of the need to manage food waste.

[8] The paper by Pietzsch et al. includes a detailed annotated bibliography which, in the author's opinion, will prove valuable to graduate students and researchers.

1.1.4 Brea le Air

As noted vater is crucial to human survival as is atmospheric dry air but, should be clean and drinkable from a safely managed source and ideally, pollutant free and contain a sufficient amount of oxygen, between the ai 23%[9] [27, 90]. That is not to say that drinking water and breathing air met international standards will result in abrupt death but, depending ab lity, there will be health effects from mild to serious and these can y to an individual's premature demise [91]. The detrimental health and l effects of polluted air have a long-documented history with count- s of legislation being established to counter the problems, but these enforced and it was only in the latter part of the twentieth century that s was taken to combat air pollution, especially in the United States [92]. anding all the efforts to clean the air, almost 7 million deaths annually are to air pollution and the World Health Organization (WHO) claim that 99% obal population in live in places where pollution exceeds their air-quality ies [93, 94]. This alarming statistic does not mean that 99% of people live work in situations where heavily polluted air is ever present, but that at sometime over the period of the year almost all people will experience breathing contaminant laden air. The challenge is to reduce air pollution, not waste clean air.

1.1.5 Freshwater and Its Uses

Although the Earth is mainly water rather than land only a very small proportion is fresh-water, about 3%, of which a sixth is readily available. The main sources of fresh water are 'groundwater, i.e., water which comes from precipitation and infiltrates into the ground forming 'aquifers', and 'surface-water runoff' which, as the term implies, is water which runs off the land mainly under the action of gravity [95]. These sources are part of Earth's natural water-cycle which is fundamentally underpinned by the global rates of precipitation, condensation, and evaporation that together provide the continuous circulation of the sensibly fixed amount of water in the Earth-Atmosphere system[10] [96]. Not only is drinking water vital to human survival but, since 2010, it has been recognized by the UN General Assembly as a human right [97]. So, if the amount of global water is fixed why is drinking water an SDG target? First of all, as the amount of water is finite and constant then as the global population rises the amount of water per individual is bound to decrease. But, secondly, since safe drinking water in the modern world has to be produced from fresh water sources, sometimes recycled wastewater, and, on occasion, from the desalination of sea water, the amount that has to be generated will likely have to increase at a sustainable level.

---

[9] By volume/mole fraction.

[10] This description of the water-cycle is a simplified explanation of what in reality is a complex system.

1.1.4 Breathable Air

As noted earlier, water is crucial to human survival as is atmospheric dry air but, ideally, the water should be clean and drinkable from a safely managed source and the air should be pollutant free and contain a sufficient amount of oxygen, between about 19.5% and 23%[9] [27, 90]. That is not to say that drinking water and breathing air which does not met international standards will result in abrupt death but, depending upon their quality, there will be health effects from mild to serious and these can lead eventually to an individual's premature demise [91]. The detrimental health and environmental effects of polluted air have a long-documented history with countless examples of legislation being established to counter the problems, but these were rarely enforced and it was only in the latter part of the twentieth century that firm actions was taken to combat air pollution, especially in the United States [92]. Notwithstanding all the efforts to clean the air, almost 7 million deaths annually are attributed to air pollution and the World Health Organization (WHO) claim that 99% of the global population in live in places where pollution exceeds their air-quality guidelines [93, 94]. This alarming statistic does not mean that 99% of people live and work in situations where heavily polluted air is ever present, but that at sometime over the period of the year almost all people will experience breathing contaminant laden air. The challenge is to reduce air pollution, not waste clean air.

1.1.5 Freshwater and Its Uses

Although the Earth is mainly water rather than land only a very small proportion is fresh-water, about 3%, of which a sixth is readily available. The main sources of fresh water are 'groundwater, i.e., water which comes from precipitation and infiltrates into the ground forming 'aquifers', and 'surface-water runoff' which, as the term implies, is water which runs off the land mainly under the action of gravity [95]. These sources are part of Earth's natural water-cycle which is fundamentally underpinned by the global rates of precipitation, condensation, and evaporation that together provide the continuous circulation of the sensibly fixed amount of water in the Earth-Atmosphere system[10] [96]. Not only is drinking water vital to human survival but, since 2010, it has been recognized by the UN General Assembly as a human right [97]. So, if the amount of global water is fixed why is drinking water an SDG target? First of all, as the amount of water is finite and constant then as the global population rises the amount of water per individual is bound to decrease. But, secondly, since safe drinking water in the modern world has to be produced from fresh water sources, sometimes recycled wastewater, and, on occasion, from the desalination of sea water, the amount that has to be generated will likely have to increase at a sustainable level.

---

[9] By volume/mole fraction.

[10] This description of the water-cycle is a simplified explanation of what in reality is a complex system.

Hence, the rationale for the SDG target 6.1, i.e., "*Drinking Water: By 2030 achieve universal and equitable access to safe and affordable drinking water for all,*" and to monitor the progress towards this target a single indicator, 6.1.1, is used which is a measure of the "*Proportion of population using safely managed drinking water services*" [98].

As with all the SDG targets and indicators there are so-called 'custodians' who are responsible for compiling, verifying, and reporting the data associated with a particular indicator or indicators [99]. For drinking water, the custodians are the WHO and the United Nations Children's Fund (UNICEF) with another UN Department, UN Water, overseeing the processes and their timelines for the data compilations [100]. As reducing waste is considered to be an essential factor in addressing both climate change concerns and achieving the transforming our world agenda, the data from the relevant SDGs provides the insights into what actions need to happen. However, as with many SDG targets and indicators, care needs to be taken in the interpretation of the terminology, for example, in the case of drinking water, the assessment of 'safe drinking water' and 'safely managed drinking water' may appear to be equivalent but they are not synonymous. The first term 'safe' defines the water source and quality whereas the second term 'safely managed' defines the 'level of service' [27, 101]. Moreover, in both cases, the drinking water must come from an 'improved' source, essentially one where human excreta and chemical contamination has been hygienically separated from the water to be supplied by a public standpipe, a piped household water connection or one of several other 'protected' facilities. Such systems normally enable the delivery of safe drinking water although there could be times when the water becomes contaminated. In such situations the provider or a local health authority issues warnings, e.g., a 'boil water' or a 'do not use' notice.

In 2020, the WHO reported that 93% of the global population had access to an improved water source, so it could appear that a modest reduction in water waste would be sufficient to meet the universal drinking water targets [102]. However, the same annual dataset used by another authoritative source revealed that only about 75% of the population, 5.8 billion people, had access to safely managed drinking water, stating that "*One-in-four people do not have access to safe drinking water*" [27]. This suggests that there is a far more significant challenge if the SDG targets are to be met. These apparent contradictions are the result of the use of terminology, the difference in quoted percentages being due to the definition of service levels as summarized in Table 1.3 together with the 2020 data for the different levels.

Trying to reconcile the data in Table 1.3 with the WHO's 2022 claim that at least 2 billion people use a drinking water source contaminated with faeces, suggests that only safely managed drinking water is fit for human consumption [102]. If this is really the case then any attempts to reduce drinking water waste should be focused on households and commercial enterprises such as, but not limited to, restaurants, supermarkets, food producers and hotels together with providers of municipal water suppliers. For all these sectors, national authorities, such as the USEPA, have long promoted water conservation and the efficient use of water [103–105]. The food industry is heavily regulated in many countries with respect to the quality of water used in processing post-harvest products and analyses of the volume of water required

Table 1.3 Drinking water service level definitions with 2020 data [101, 102]

Drinking water service level	Definition	2020 Population (Millions)
Safely managed—improved	Located on premises, available when needed, free from contamination	5,800
Basic—improved	Collected from a source located within 30 min for a round-trip including lining up and waiting	1,200
Limited—improved	Collected from a source location requiring more than 30 min for a round-trip including lining up and waiting	282
Unimproved—unprotected	Water taken from unprotected dug wells and springs	368
Surface—untreated	Water collected directly from untreated surface water or a river, dam, lake, stream, canal	122

to produce common foodstuffs have lead to numerous calls for some populations to change their diets, e.g., from mainly meat to more fruits and vegetables [106, 107]. In addition to the overtures that such dietary changes will providing healthier nourishment and be positive contributors to climate change mitigation, the production processes for certain foods are less water intensive than others, as measured by 'Virtual Water' quantities. The latter is the hidden quantity of water that is used in the production of goods of which foodstuffs is but one example [108–110].

Saving water by dietary changes may be appealing to some, but for many it could require difficult cultural transformations. Moreover, the water savings gained by selecting particular foods would need to be contrasted with their nutritional energy values,[11] so, for example, while per kilogram cabbage only consumes just over 5% of virtual water compared to the same weight of chicken meat, to obtain the same food energy almost 9 times more cabbage than chicken would need to be consumed, as illustrated in the virtual water-food energy exemplars given in Table 1.4, with the final column showing the amount of food which would need to be consumed to meet a typical daily calorie intake.

Diets aside, the production of almost all foodstuffs (except for crop irrigation) requires freshwater and the quality of that water is frequently the same standard as human drinking water, and, in some cases, even more stringent [112, 113]. Yet, the regulations for livestock drinking water and water use in the food processing industry regulations are not as frequently monitored and enforced in comparison with safely managed and improved human drinking water [114, 115]. Despite the emphasis on drinking water for humans in the SDGs, worldwide, only about 11% of 'withdrawn' freshwater is used to provide municipal drinking water, albeit, there are wide variations by region and country [116, 117]. However, care must be taken when

[11] Measured in Calories or Kilocalories which are interchangeable units for food energy.

Table 1.4 Virtual water consumption and food energy exemplars [108, 110, 111]

Foodstuff	Virtual water consumption, L/kg	Calories/kg	Water consumption litres/calorie	Amount in kg for daily 2,500 cal
Solids				
Chocolate	17,196	4,480	3.84	0.56
Beef	15,415	1,710	9.01	1.46
Chicken	4,325	2,190	1.97	1.14
Apple	822	520	1.58	4.81
Banana	790	890	0.89	2.81
Cabbage	237	250	0.95	10
Liquids				
	Litres/litre	Calories/litre		Litres for daily 2,500 cal
Milk	1020	600	1.7	4.17
Wine	436	800	0.55	3.13
Beer	296	430	0.69	5.81

interpreting 'water-use' data since the amounts withdrawn[12] from freshwater surface and groundwater sources and the amounts consumed are not the identical [118]. For example, in the United States, the largest proportion of freshwater withdrawals are for utility scale thermal power plants, about 41%, but only approximately 1% is consumed depending upon the cooling system used [119, 120].

Increased efforts are being made to address the issues associated with somewhat ill-defined water use terminologies, albeit, new terms such as 'renewable' are now being applied to groundwater resources when extraction from deep aquifers becomes technically challenging or too expensive, or both [121, 122]. Nevertheless, over the past decade improvements have been made in the precision of the methodologies used in the measurement of freshwater withdrawals and consumption, particularly in the EU [123]. These, together with more rigorous clarifications of terms used to describe the measurements, have not yet enabled broader global datasets to be established other than the categorization of water withdrawals into 3 sectors, agriculture, industry and municipal. Even so, the municipal portion may also include some agricultural and industrial use if the associated facilities are connected to a public water distribution supply network [116]. Why are these data important? The simple answer is that if any of the precious resource of freshwater is not to be wasted it is vital to know where it is being used and the relevant proportions. From the data that are readily available, except for Europe, the main use of freshwater is for agriculture, although, as shown in Fig. 1.5, there are significant regional variations [117, 124]. It can also be seen in Fig. 1.5 that the freshwater withdrawals used for municipal purposes, including the drinking water related to SDG 6.1, are a relatively small proportion of total water

[12] Water withdrawals are referred to as 'abstractions' by some national agencies.

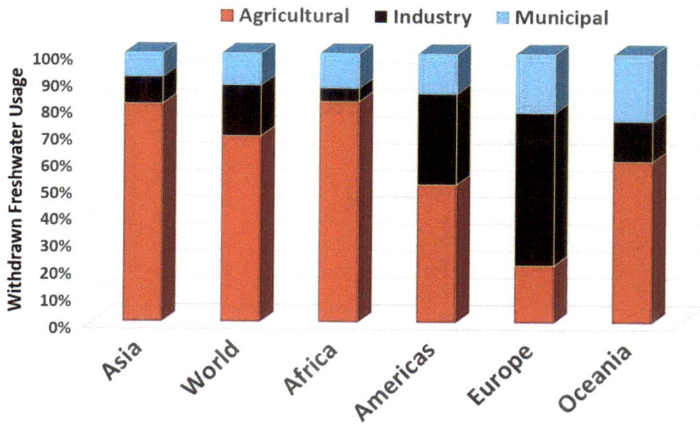

Fig. 1.5 Freshwater withdrawals proportions by region and sector [117, 124]

withdrawals of the 3 'core' sectors. Thus, when focusing on reducing freshwater waste is municipal water where the efforts should be concentrated?

The regional data as presented as Fig. 1.5, while helpful could be misconstrued, as not only are there measurable regional variations but, within regions, the municipal withdrawals by individual countries can be vastly different, Fig. 1.6 [116]. A typical example is the African region where overall the percentage of total water withdrawals for municipal water is 13%, similar to the world average, but the differences between the Democratic Republic of Congo, Botswana, and South Africa are pronounced, in 2017 being 68%, 52% and 20% respectively [116]. There are numerous reasons for such variances not least of which are geographical location, annual precipitation levels, and a particular country's economic fundamentals. Evidently then, there is no singular universal action that can be taken to reduce freshwater wastes. However, as illustrated in Fig. 1.5, apart from Europe, global freshwater withdrawal is dominated by the agricultural sector and this situation has been forecast to be further exacerbated by population growth and the need for more food. Indeed, it has been determined, that by 2050 global food production will require three-times the amount of water it used in 2013 [110]. Should this prove to be accurate the competition for freshwater resources will become relentless especially since, according to the UN, the land and water requirements for agriculture and food production are already strained [125].

The data presented in Figs. 1.5 and 1.6 shows only the freshwater that is withdrawn in the core sectors but, as previously mentioned using the United States statistics as an example, thermal power plants actually account for the greatest volumes of extracted freshwater. Is this situation peculiar to the United States? The answer is no, but the global data are somewhat limited. However, the EU, which has the world's 2nd largest economy based on Gross Domestic Product (GDP) and a population over a third larger than the United States, has been collecting and collating such data since at least 2010 [123]. In fact, the EU has identified and defined 7 economic sectors for the determination of freshwater withdrawal including 'electricity cooling'. In 2019, the

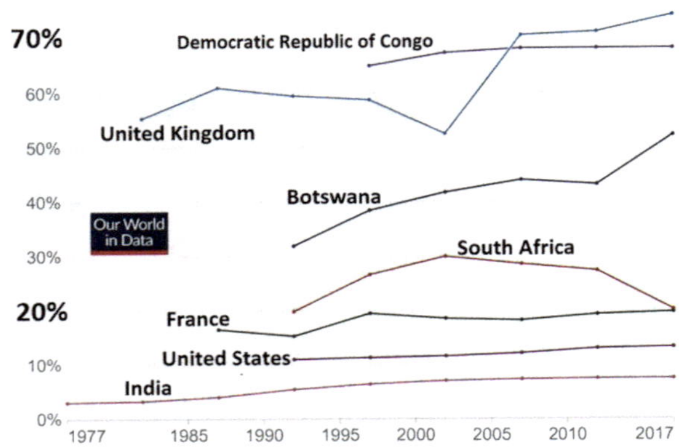

Fig. 1.6 Share of freshwater withdrawals for municipal use—exemplars [116]

cooling requirements for electricity generation represented 32% of all withdrawals, the largest single sector, a trend that has been the case for almost 20 years, as shown in Fig. 1.7. Notwithstanding the use of a 3 or 7 sector approach to determine the reasons freshwater is withdrawn, and in what amounts, to ascertain potential areas of waste reduction, it is surely the consumption data that are important. This has long been recognized, especially for the agricultural sector, with published records in the United States, for example, dating back over a century [126]. It is not surprising that this sector has been the focus of efforts to determine consumption rates since, although the new methodologies indicate that thermal power plant withdrawal needs can be higher, the actual consumption data that is available shows that agriculture is, by far, the largest consumer of freshwater [127].

Determining the actual consumption of freshwater by humans and human activities has proved, and continues to prove, a monumental task, not only on a national and global basis, but also at a local level particularly since seasonal, geographical, and economic factors all play crucial roles, as well as sufficient in-situ infrastructure to measure consumption rates in the situations where waste can occur, e.g., agricultural irrigation. Attempts at modelling consumption amounts based on withdrawal data and the use of semi-empirical consumptive coefficients (CU) have experienced varying degrees of success [128, 129]. There has been a growing realization that water

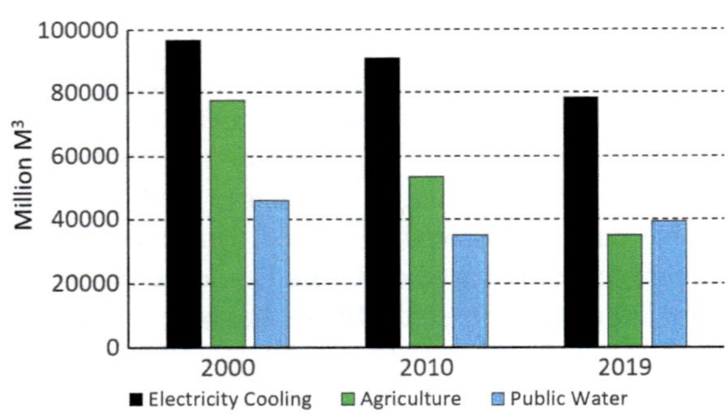

Fig. 1.7 European freshwater withdrawals trends 2000–2019 [123]

is not a 'given commodity' but a crucial and economic resource with environmental implications for all. The UN have developed a framework consisting of an extensive set of guidelines and recommendations for the reporting and accounting of water statistics as part of their system of Environmental-Economic Accounting (SEEA) [130]. Until the economic factors were embodied in water use analyses the common use of a 'water-use efficiency' (WUE) measure determined how much product, i.e., food, goods, and services, was obtained per unit volume of water used [131].

With the advent of the SEEA-water framework and the adoption of the transforming our world agenda, specifically SDG target 6.4,

> By 2030, substantially increase water-use efficiency across all sectors and ensure sustainable withdrawals and supply of freshwater to address water scarcity[13] and substantially reduce the number of people suffering from water scarcity,

and indicator 6.4.1, "*Change in water-use efficiency over time*", a modified WUE is being used where, rather than the amount of 'product' being measured, it is the product's value in US dollars per unit water volume that is one of the target's key indicators of progress [132, 133]. This measure is as much about the economic growth of a country as it is about access and use of water, but it is a way, albeit somewhat obfuscated, of assessing if water is being wasted or misused. The UN's perspective is that, if the 6.4.1 indicator value increases at a greater pace than a country's economy then, in essence, water is being used wisely and in a sustainable manner, whereas if the indicator grows at a slower pace than the economy, that country's opportunity for sustainability could be jeopardized and it is highly likely that it will encounter acute freshwater water resource availability and access challenges [133]. Most certainly, great efforts have been made to collect and collate data pertinent to this particular indicator, including the provision for participants of e-learning modules on how to apply the adopted methodologies to measure the many variables involved in the processes and highlight "*the impact that monitoring results may have on national decision-making*" [134]. Nevertheless, the interpretation of such results is not as straightforward as may be envisaged, for example, a subscription-based website for lawyers lists at least 7 different definitions of 'consumptive water use' that could be employed or encountered in legal documentation [135].

There is clearly more to reducing wasted freshwater than fixing household leaks, albeit, it has been estimated in the United States that the amount of drinking water wasted through leaks could provide 11 million homes with all their annual water needs [136]. While this may seem a trivial amount it would be sufficient to supply the household[14] water needs of whole countries such as Canada, Poland, or Iraq. Yet, in global terms, household water waste is trifling in comparison with the amounts wasted in agricultural irrigation systems [137]. Despite only a very small percentage of the global water resources being freshwater there is a sufficient amount to meet the needs of the world's population, but unfortunately it is not evenly distributed geographically, and even where the supplies are more than sufficient it is not always managed

[13] When normal demand exceeds available supply.

[14] Based on 4 occupants per home.

sustainably, either annually or during particular climatic seasons. These situations usually occur in low-income countries [116]. In all countries there is persuasive evidence that too much water is wasted or polluted and, in many instances, because of the so-called 'yuck-factor', there is a public reluctance to use 'treated' wastewater, although 80–90% of municipal water, especially from residences, becomes treatable wastewater [138, 139]. However, the results of recent surveys conducted in the Netherlands, the United Kingdom, and Spain, although limited in sample size, have indicated that the aversion is far less than has been presumed, historically [140].

Adequate and safe supplies of freshwater are essential for the survival of the planet. It is vital for people and wildlife and almost every living thing. It is used for food production, power generation, manufacturing, construction, sanitation, and an endless list of other human activities. Yet, the natural supply, 'the water-cycle', is finite and its use has to be reconciled with population growth, at least for much of the remainder of the twenty-first century. The world can ill-afford to waste water, although most of society continues to use it inefficiently and uneconomically. This is an unsustainable scenario. Consequently, efforts to quantify and enumerate freshwater uses have been accelerating as the world seeks to establish a sustainable future. These efforts have encountered numerous technical and economic challenges coupled with societal road-blocks, cultural and psychological, as discussed briefly in the previous few paragraphs. A more in-depth discussion on the uses and abuses of the global freshwater resources is outside the scope of this chapter and indeed would need an encyclopedic approach to the subject matter. Nevertheless, it is hoped that readers have gained adequate insights into the issue.

1.1.6 Continuing Focus on Food

In the opening section and sub-sections, an attempt has been made to provide some insights into the types of wastes encountered in everyday life. To discuss, even in the briefest of detail, the whys, and wherefores of all the categories and classifications of waste, Table 1.1, and their reduction is outside the author's selected range of this chapter. However, in the next sections, while the focus of waste reduction is maintained, a particular emphasis is placed on a crucial aspect of the w-e-f nexus, i.e., food. As previously highlighted, the food waste in one country could feed the undernourished in other countries and regions. Numerically this may be so, but the implementation of the concept would encounter a multitude of logistical as well as geopolitical obstacles. Globally, about 11% of food waste is generated by households, but there are wide variations between countries and regions. However, more is wasted (17%) in the earlier part of the food supply chain between harvesting and being available to retail consumers, referred to in the SDG literature as 'farm to fork', and again there are wide global variations [141, 142]. To distinguish food wasted once it arrives at the retailer, provider, and consumer from that wasted before it does so, the commonly used terms are 'food waste' and 'food loss', respectively [143]. The Food and Agriculture Organization of the United Nations (UNFAO) has suggested descriptions for food loss and waste, which they refer to as collective as FLW, namely: [144],

> Food loss: the decrease in the quantity or quality of food resulting from decisions and actions by food supply chain actors from the production stage up to, but excluding, retailers, food service providers and consumers.
>
> Food waste: the decrease in the quantity or quality of food resulting from decisions and actions by retailers, food service providers and consumers.

These definitions form the basis of some of the indicators used to measure the progress towards achieving SDG goal 12 [7]. Although the term 'waste' rarely appears in the UN's 'Transforming Our World' document it would be unfair to assert that waste has been neglected in this plan to achieve a sustainable future. In particular, SDG 12, with the goal to "*Ensure sustainable consumption and production patterns,*" has a number of targets specifically aimed at solving the global waste problems, such as target 12.5 "{to} *substantially reduce waste generation through prevention, reduction, recycling and reuse,*" but only one target, 12.3, has a stated numerical objective, i.e., " {to} *halve per capita global food waste at the retail and consumer levels and reduce food losses along production and supply chains, including post-harvest losses*" [7]. The UNFAO definitions of food loss and food waste given at the end of the preceding paragraph provide the basis for a common approach to the measures, i.e., indicators, needed to assess how well national and international efforts are doing in accomplishing the 'food' targets. As with many of the indicators and sub-indicators of the SDG targets, numerical indexes have been universally developed to provide a readily understandable portrayal of the efficacy of the adopted strategies. In the case of SDG 12.3, these indexes are the Food Loss Index (FLI)[15] and the Food Waste Index (FWI). Although simple in concept the methodologies used to determine the values of FLI and FWI are complicated, and they are a 'work-in-progress' [145]. Even so, these indexes are providing noteworthy insights into the issues encompassing the challenges of feeding the world [16, 146]. Food Waste and Food Loss issues pertinent to waste reduction are discussed further in Sect. 1.3.

However, before discussing aspects of food waste generation and the handling of MSW, as previously alluded to (Sect. 1.1.2), responses to the queries raised about how national economies and population size impact waste generation and handling (Sect. 1.1.3) are provided.

1.2 Waste, Income and Population

The purpose of this section is to explore the background to how national economies and population size impact waste generation and handling to gain some understanding of widely held perceptions about waste production, i.e., are they instinctive

[15] A logical acronym, but if used in a literature web-search it can overlap with other uses such as the medical profession's Fatty Liver Index; and the same is true for the occurrences of the FWI acronym.

or evidence based. For example, on a per capita basis it is frequently presumed that those living in a high-income economy generate more waste than those inhabiting lower-income countries and regions, and the richer nations are able to afford the infrastructure to handle and treat waste. Moreover, it is supposed that countries with larger populations generate more waste than those with fewer people but similar incomes. This would seem to be an obvious state of affairs since, if the income per capita of two countries is the same then, if the rule-of-thumb maxim about waste generation being proportional to income is accurate, the country with the larger population will generate more waste. Here are some examples.

Two countries with almost identical per capita incomes are France and the United Kingdom (UK), but the former generates about 10% more waste, although the UK has a slightly larger population (approximately 4%) [147–150]. In both cases, the major source of waste is the construction industry [151]. Mexico and Japan are two countries with almost identical populations, within 2%, but Mexico generates about 10% more waste than Japan. Is this because Mexico is a richer country in terms of income? The answer is no. Indeed, the reverse is true, because the per capita income in Japan is almost 4 1/2 greater than Mexico [150]. On the other hand, the Republic of Türkiye (Turkey) which has a population almost the same as Germany's but an income level of about 1/5, produces over 30% less waste. The bulk of Germany's waste comes from the construction industry whereas Türkiye produced almost no construction waste[16] [151].

These examples suggest that the population size, not surprisingly, does play a role in waste generation but income levels can also impact the amounts generated, to varying degrees, resulting in increased generation in some instances and reduced generation in others. An arguably persuasive example of such income effects is that although China has a population some 4 1/3 greater than the Unites States they both produce approximately the same annual amounts of MSW, suggesting that the 1:6 per capita income ratio in favour of the Unites States is the more dominant factor [16, 150].

Therefore, since national incomes levels cannot, or at least should not, be neglected in the assessments of waste generation, it is typical for much of the publicly available waste data to be presented against a background of national or regional per capita incomes [16]. These income levels are defined by the World Bank Group (WBG) who classify almost 200 countries, with populations greater than 30,000, into 4 groups, low-income being $1,085 USD[17] or less, lower middle-income is $1,086 to $4,255, upper middle-income is $4,256 to $13,205, and high income is $13,206 or greater [152]. These income levels may appear somewhat low, but it has to be remembered that they are per individual and not per household. Nevertheless, while the World Bank likely has good reason to define the levels and ranges within these income bands, they can create a somewhat perplexing portrayal to the lay-person. For instance, the ratio between the bottom of the lower middle-income range and

[16] For many years prior to the devastating earthquake of February 2023 when at least 1.5 million people were homeless.

[17] USD = United States Dollar.

the top is 3.1, for the upper middle-income bracket the ratio is 3.9, and for the high-income range it is 9.3. Thus, using the World Bank's 'Atlas' method of calculating the Gross national income (GNI) per capita, Seychelles, with a population of just over 100,000, has a per capita income of $14,540 and is classed in the same income bracket as Bermuda, population approximately 62,000, whose per capita income is $122,470 [150]. Is it reasonable to compare these two high-income countries in terms of waste generation? Perhaps, but as waste data are only available for Bermuda this is not possible. Interestingly, although the difference in income between the two is dramatic, around a quarter of the population in both cases is considered to live below their national poverty line, which just over $44,000 per annum in Bermuda and $4,000 in the Seychelles. Is the proportion of people living below a defined poverty line as important an economic factor in the reduction of waste as income?

In 2019, less than 9% of the global population lived in extreme poverty, down from almost 38% three decades earlier (Fig. 1.1), but this still meant that 659 million people, almost 10 times the population of the UK, were living below the international poverty line [11, 153]. Those living in such conditions are likely to be paltry contributors to waste generation, other than normal human effluent, but if they are lifted out of poverty then presumably, they would generate other forms of waste and so heighten the challenge to achieve global zero-waste? This could pose ethical dilemmas for a sustainable world given that the key goal of the UN's sustainable development agenda is SDG 1, i.e., "*To end poverty in all its forms everywhere*" [7, 98]. This is an ambitious goal, albeit the 7 SDG 1 targets and 13 indicators are not as prescient. Examples of this 'realism' include indicator SDG 1.1 which states that, "*By 2030, eradicate extreme poverty for all people everywhere*," the measure to be an international poverty line, and target 1.2 which is to reduce by at least half the proportion of people living in poverty according to "*national definitions*" [98]. As with the global income levels, the World Bank defines the international poverty level based on the value of goods needed to sustain one adult per day and is periodically updated [154].

The 'goods' in question are basic food requirements, clothing, and shelter but the values or costs of water, energy and sanitation are not included which is perhaps unfortunate as the former is a necessity for survival and the latter two impact health, well-being, safety, energy, and sanitation, i.e., the quality of life [154, 155]. The most recent update (2022) to the international poverty line, which defines those living in extreme poverty, is $2.15 per day per individual [156]. This level has increased since 2019 and now slightly over 9% (719 million) of the global population lives in extreme poverty but it should be noted that, once again, the methodologies for calculating the poverty line have also been updated [155, 156]. There is a reasonable correlation between national per capita and the level of its respective poverty line, but not necessarily between poverty levels and the share of the population living below the level. In Ethiopia, for example, with a per capita annual income of just $940, 23% of the population live below the poverty line of $2.04 per day, whereas Bangladesh, with an income close to 4 times that of Ethiopia but with a poverty line only 5% greater, has the same percentage share living in poverty. Conversely, the United States which has an income level more than 7 times than of Türkiye has a lower portion of its population living below its poverty line, but in comparison with Vietnam whose income level is about one-twentieth of the United States, the portion living in poverty is 60% higher. The variations between World Bank defined income classification and poverty levels are shown on Table 1.5. If the United States' own definition of its poverty level were to be used than all people living in low, lower medium-income, and upper medium-income ranges would be labelled as living in poverty.

Overall, it does seem that, in many instances, the assertions made in the opening paragraph of this sub-section have some merit but there are ample examples to show that population size and income levels alone, including poverty lines, are insufficient to provide uncomplicated insights into effective ways of reducing waste. To some extent this is also indicative of the types of waste reduction approaches being associated with specific regions and countries rather than there being a 'one-solution-fits-all', as already illustrated when dealing with freshwater use, see Fig. 1.5. These rather convoluted situations are invariably encountered when all forms of waste, such as those identified in Table 1.1, are considered. There are numerous examples of this global mishmash approach to waste generation. High-income countries account for about 16% of the global population yet generate 34% of the municipal waste and almost 90% of industrial waste, but send less of their waste (about 39%) to landfill sites, compared with upper-middle income countries [16]. This high level of waste generation can be explained by the industrial economies of such countries, but as they tend to be more urbanized, i.e., 80 + % living in cities, why is not more waste disposed of in landfills [16, 158]? Could it be that high-income countries recycle more of their waste, particularly MSW, than the other income groups? North America recycles over a third of its waste, but the other income groups only recycle small amounts, between 3.7 and 6% [16]. Could measurably increased recycling then be the answer to waste reduction for 84% of the global population? Once again, in attempting to delineate the causes of waste generation using statistical data and instinctive correlations, another rabbit-hole is encountered! Obviously addition factors other than population size, income size and poverty levels also influence waste generation, and some of these have already been encountered in Sect. 1.1.2, i.e., the composting, recycling, and

Table 1.5 Income classifications and their respective poverty lines [152, 156, 157]

Income classification	Income range $USD/ capita per year	Median poverty line $USD/day	Median poverty line $/ bottom range$ over a year, %
Low	≤1,085	2.15[a]	72.3
Lower-middle	1,086–4,255	3.63[b]	122.0
Upper-middle	4,255–13,205	6.85[b]	58.7
High	≤13,206	24.36[c]	67.3

[a]World Bank's international poverty line for extreme poverty
[b]World Bank's analyses based on national poverty lines data
[c]The United States Government defined poverty level is $34.79

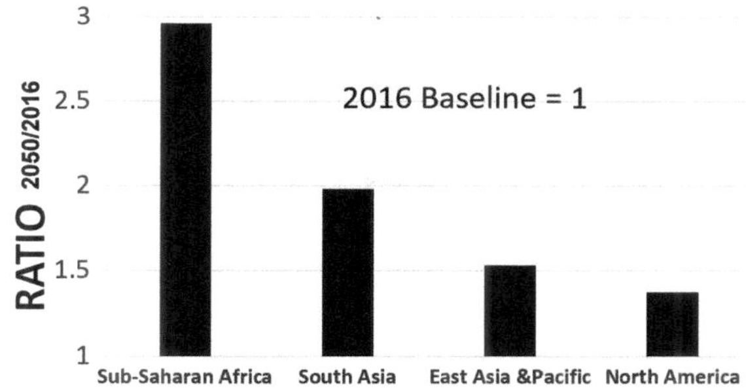

Fig. 1.8 MSW generation ratios 2050 compared to 2016 using data from [16]

waste-to-energy conversion of MSW. These could be roughly described as 'cultural' but, as discussed in the next section, there are other cultural elements that impact waste generation and management.

Regardless of the reasons for the scale of waste generation, there are concerns that the main challenges to achieving zero, or reduced, waste in the future will come from regions not usually associated with high rates of waste production and that lack the infrastructure to deal with these increases, which could be dramatic in some cases. These possible circumstances can be shown by the forecasts for MSW generation by 2050 compared to 2016 (Fig. 1.8).

While the sub-Saharan Africa region is forecast to have a rapid proportional increase in its MSW generation, despite many of its 47/48 countries having incomes in the low or low middle income range, ratios are not necessarily reflective of absolute amounts. However, the region already has a sizeable population, close to three times that of the United States, and by 2050 there could be over 5 times more people living there compared to the United States and North America including a very small global proportion of older people [86, 159]. The amount of MSW generated by sub-Saharan Africa will be 30% more that North America in 2050 whereas in 2016 it was 40% less, in millions of tonnes (Fig. 1.9). Is this change because of population increases or the demographics of such increases? The latter appears to be an under-researched area of waste generation so there is a dearth of evidence on aging population effects and a reliance on anecdotal rhetoric and speculation, such as the retired wealthy have more time to spend their money creating more waste, some retirees with mobility issues may mismanage their waste because of their inability to properly dispose of it, whilst others have lived most of their lives in the times of 'take-make-use-discard' rather than the modern sustainability trend now being adopted by society [160]. However, more substantive research on demographics and waste management is becoming available e.g., [161, 162]. Plainly, population demographics is yet another influence which must be added to the basket of factors impacting waste generation and management.

To a lesser or greater extent, the factors identified so far, population growth and dynamics coupled with economic parameters, such as income and poverty levels, effect international, regional, and national production of waste and food is no exception. But, in the case of food waste, there are also other influences such as those

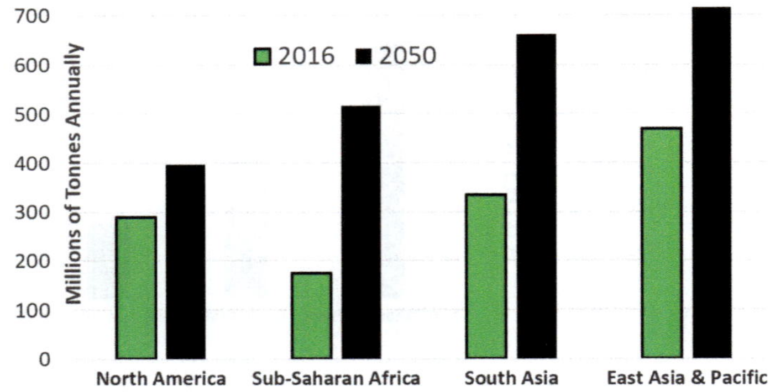

Fig. 1.9 MSW generation forecast for 2050 compared to measured in 2016 [16]

associated with attitudes, behaviour, and awareness, i.e., essentially, cultural norms [163].

1.3 Food Waste and Food Loss

Why should food waste be a concern when, according to the World Food Programme (WFP),[18] global food production is sufficient to nourish all of humanity [164]? The concerns are two-fold. First of all, with a rising global population will there be enough 'dietary energy' to meet the needs of everyone in the future? The UN's last population forecast, based on their 'medium' scenario, is for a global population of 10.4 billion by 2100, about 2.4 billion more than at present[19] [86]. Based on this projection and present global food production more food will needed but maybe not as much as envisaged. The main reason is that much of the current global food production is never eaten and if made available could meet the needs of 2 billion people [164]. On the face of it then if all produced food is consumed, i.e., no waste, then over the next 7 decades or so, production would need to increase by only 20%, surely not an unreachable goal as agricultural production, which can include non-food items such as cotton and wool, has risen by almost 400%[20] in the 6 decades since 1961 (Fig. 1.10) especially in the UN designated 'Least Developed Countries (LDC)'. These are low-income countries "*confronting severe structural impediments to sustainable development*" and "*have low levels of human assets*" [165–167]. The challenge then appears not to be the global scale of food production but combating the squandering of the food that is produced. However, this is easier said than done since,

[18] WFP is an international non-government organization (NGO) which is mainly funded by global governments and has relationships with many United Nations departments and over 1,000 other NGOs.

[19] 15 June 2023.

[20] Output based on $US value in 2015 [165].

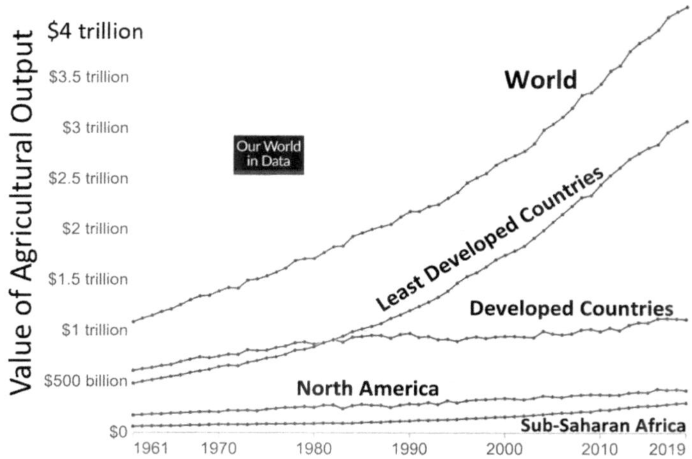

Fig. 1.10 USD financial value of agricultural output 1961–2019—exemplars [165]

lamentably, global food production is unevenly distributed by geography, a nation's land area coupled with sea and lake access, the quality of their agricultural ecologies and by the factors already identified such as population size and income level. Severe weather events and armed conflicts can also have negative effects on food production creating even more challenges for vulnerable populations, especially in sub-Saharan Africa and parts of Asia. Thus, the second WFP concern, shared by such bodies as the UN, is that the unevenness of food production causes 'chronic' hunger, referred to as the 'prevalence of undernourishment' or simply undernourishment [168].

This imbalance in global food production and the lack of regional, national or local food availability is manifest in the share of a population that is defined as 'undernourished', i.e., individuals not receiving the minimum 'dietary' energy they require to maintain a normal life including a healthy weight for their usual level of activity [76]. Measured in the calorific value of an individual's food intake, the minimum needs will vary according to an individual's demographic statistics, e.g., age, height, weight, activity and so on [169]. There will, however, be more influences on the minimum level such as occupation, e.g., a firefighter, construction worker, or professional athlete will require a greater calorie intake than someone in a more sedentary profession such as an office worker. Moreover, all these variables will have slightly different impacts depending upon where an individual lives. To take account of such numerous aspects the UNFAO uses a number of factors to determine a 'parametric probability density function' from which they calculate the minimum calorific requirements for each country [169]. Nevertheless, although painstaking efforts are made to determine the scale of undernourishment, the computed data so far are at best insightful estimates. These data can be illustrated in terms of the share of the population suffering undernourishment, but perhaps a clearer portrayal and

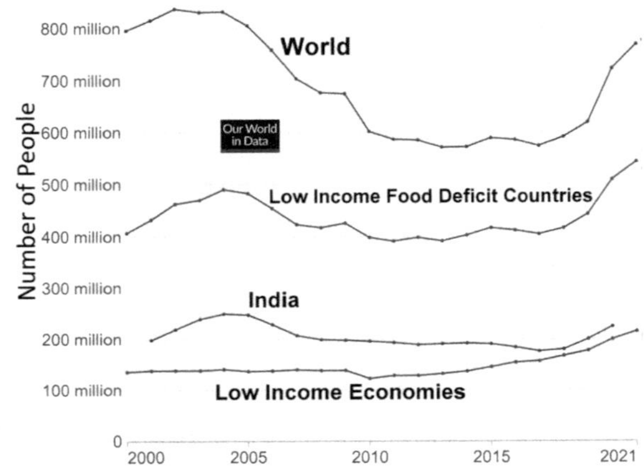

Fig. 1.11 Number of people experiencing undernourishment—exemplars [76]

appreciation of the magnitude of these issues can be gained by comparing the actual number of people experiencing such deficiencies, as shown in Fig. 1.11 [76].

It should be noted that on this chart another grouping is used in the analyses, namely 'Low Income Food Deficit Countries' (LIFDC) [170]. An LDC country is not necessarily a LIFDC, and vice versa, and the income classification of some LIFDC is not always low-income but could be low middle-income as the threshold used in designating a country a LIFDC is $1,395. A somewhat elaborate clustering system, but essentially one that identifies a country by income-level and as a net importer of basic foodstuffs, albeit that many sub-Saharan countries lack sufficient foreign exchange to purchase all the external foodstuffs necessary to nullify their levels of undernourishment. In 2021, there were over 540 million people living in LIFDCs who were undernourished, of which almost 50% lived in sub-Saharan Africa and almost 770 million, and maybe more worldwide, probably 10% of the global population, were determined to be in the same situation [76, 171]. The trend since 2019 is for more, rather than less, people globally to be facing hunger. Not a good harbinger for meeting sustainable development indicators by 2030 and a situation which seems to be accelerating since 2022 due to post-COVID inflation and the invasion of Ukraine.

Food is essential for human and animal survival and is produced in many forms and by many methods. However, not all the food produced, reaches the intended consumers, even after it has been 'harvested'. The definition of food loss as given in Sect. 1.1.6 could include food that is never harvested for a varying of reasons such as those already mentioned, destructive weather events, changing climates and armed conflicts. Their impacts are never easy to assess and even more difficult to predict, especially when they involve instances of dire episodes of extreme hunger, i.e., famines, which cause high death rates among the affected population due to starvation and hunger related diseases [172]. Even before the adoption of such recent socio-economic philosophies as sustainable development, the global death rates from occurrences of famines had drastically reduced in the past 1 1/2 centuries mainly due to aid packages from many global governments and NGOs [172]. Notwithstanding this laudable trend, if advanced warning of potential situations of acute food shortages can be made then the likelihood of a timely emergency response, particularly from the major food exporters such as the United States, averting a famine crisis becomes a strong possibility. To this end, since the early part of this century, the UNFAO has worked with a number of international partners to develop the methodologies for an 'Integrated Food Security Phase Classification (IPC)' to inform emergency responders of where to expect acute hunger situations [173]. An early forerunner of the IPC global organization was the 'Famine Early Warning System Network' (FEWS NET) established almost 4 decades ago by the United States Agency for International Development (USAID) which was formed in response to famines in parts of Africa [174]. Today FEWS NET is an international NGO that uses the IPC methodologies to project future near-term, i.e., 4 months, and medium-term, i.e., 8 months, most likely occurrences of acute food shortages [175].

There is then a wealth of global data on food production, hunger and undernourishment, deficiencies in national food supplies, persistent and temporary, and international emergency response systems to combat actual and potential famines. These together provide a credibly wide-ranging representation of how successful humanity has been in ensuring everyone has adequate dietary energy to not only survive but lead a 'normal-life'. Arguably, since 90% of the largest global population ever known usually do not suffer from hunger, albeit that some 40% cannot afford what experts would describe as a healthy diet, anthropogenic performance in supplying food needs is worthy of an overall high grade. However, percentages are mathematical artifacts not living people and the fact that hundreds of millions of people are undernourished, especially children, must inevitably be the driving force in eliminating world hunger, complacency not being an option. So, if pursuing global reductions in food waste and food loss can be instrumental in achieving adequate food-for-all then they should be vigorously tackled. But How? In the first instance it would be useful to know who wastes the most post-harvest food. The UNEP are attempting to quantify this factor by defining and measuring, or collecting, national data for 3 sectors; namely, households, food services, and retail [176]. This is proving a daunting task, especially when collecting data from countries that are not considered high-income. The analyses to date indicates that about 66% of all food waste is generated by all income category households. Moreover, lower-middle income and low-income countries produce the most food waste per capita, perhaps not an intuitive finding. It has also been found that there are wide variations within regions, countries, and income-level clusters, the United States household food waste proportion is only about 43%, for example [177].

Most of the household food waste is deposited in 'trash' or MSW, although there are instances, mainly in high-income countries, where local authorities or agencies collect it as a separate curbside recycling material in the same way as paper, plastics and metals, and other organic materials, such as yard wastes, which are sometimes referred to as biological or green waste and are rich in nitrogen. The same system is also used for the collection of similar wastes generated by food service providers and other commercial enterprises [178]. For food waste, the processes are described as a

'Food Recovery Hierarchy' and, although there are slight differences in depictions from country to country, the one shown as Fig. 1.12 is typical [179]. These innovations are likely to be responsible for the share of food and green waste in MSW in high-income countries, approximately 32%, being significantly less than in other countries with lower incomes, where the share is 53% to 56% [16]. Clearly income level plays an important role in food waste recycling but is this situation reflected in the amount of food waste per capita? Exemplars, given in Table 1.6, for different income and population size countries along with their respective poverty lines paint a somewhat muddled picture.

The array of data being collated on food waste and food loss is used by the UN's selected custodians of SDG 12 to determine the waste index (FWI) and loss index (FLI), as discussed in Sect. 1.1.6, but as yet the employed methodologies are not wholly mature and training courses are offered to those participating in the collection, collation and analysis processes in the hope of improving data quality confidence levels [145, 180]. Nonetheless, although the calculation of the indexes, to include all UN member states, remains challenging, sufficient information is available to determine measures, for example, of food loss in certain global regions as shown in Fig. 1.13 [180]. The diagram indicates that the regions with high-income countries (mainly) lose around the global average percentage of the physical amount of food loss compared to that produced, but the low income, and largely LIFDC, countries of sub-Saharan Africa lose far more of their food production before it reaches the consumers and retailers. Maybe global help to reduce food loss in sub-Saharan Africa could have a more beneficial impact on undernourishment than efforts to reduce food waste in some other regions? Yet, here again, the situation is complicated. China is the world's largest agricultural producer but, in 2020, had to import 23% of its food needs making it the largest global food importer while, at the same time, somewhat ironically, generating almost 90 million tonnes of food waste [166]. It could be speculated that China may choose to reduce its own waste to reclaim its food self-sufficiency status of two decades ago rather than combat the food loss issues in other regions. On the other hand, the United States as the world's largest exporter

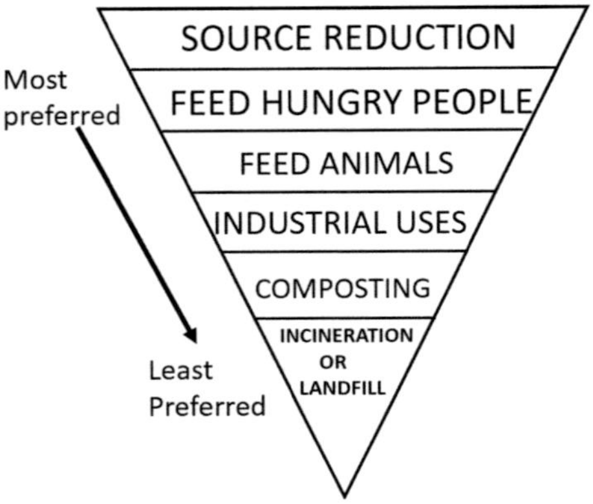

Fig. 1.12 Typical food waste hierarchy modified from USEPA [179]

Table 1.6 Economic factors and per capita household food waste based on [16] and world bank databases

Country/ region	Annual household food waste kg/capita	Total households thousands of tonnes/year	Gross national income/capita classification[a]	GNI/capita $USD[a]	Poverty line $USD annually
Austria	39	349	High	52,760	7,100
United States	50	19,360	High	70,390	12,880
Australia	102	2,560	High	57,170	12,250
Greece	142	1,480	High	20,000	5,240
South Africa	40	2,329	Medium	6,530	656[b]
China	64	91,650	Medium	11,880	1,120[b]
India	50	68,760	Low	2,150	1,130
Ethiopia	92	10,327	Low	940	745
Nigeria	189	37,940	Low	2,080	210

[a]Based on World Bank Atlas Data
[b]Medium now separated in lower and upper so the data is approximately mid-range

of agricultural products may be able to export more by reducing its food waste and loss, albeit it has a very efficient agricultural industry [166]. Regrettably, LIFDCs, by their very designation, cannot afford international market-priced food [170].

Even though some food waste is separated during collection, presuming there are such regular and permanent waste management provisions, the bulk of the waste is disposed of in landfill sites of various quality (see Sect. 1.5) or simply dumped. The lack of landfill facilities is especially prevalent in the South Asia region as well as

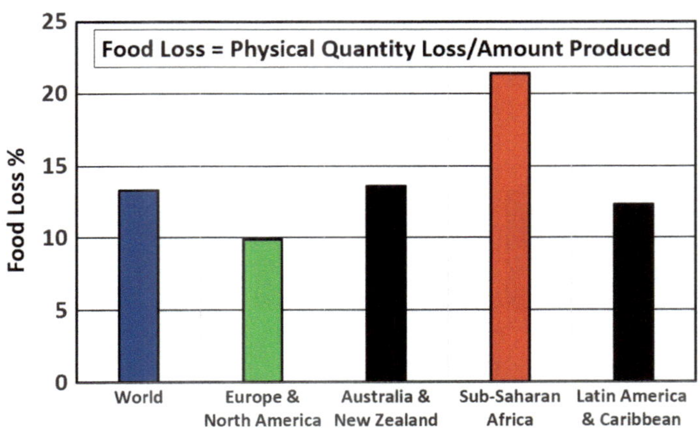

Fig. 1.13 Exemplars of global and regional food loss % [180]

Sub-Saharan Africa [16]. Waste recycling infrastructure is also largely non-existent in these regions. The scale of these waste management inadequacies is plainly visible in Fig. 1.14 [16]. These problems, and those already identified and discussed, are indicative of the myriad and complex issues involved in reducing both food waste and loss. There appears to be no single 'silver-bullet' solution to the numerous challenges, albeit some argue that there are some simple solutions [181]. But who will assume the responsibility for seeking a suite of customized local, national, and global solutions? Outside of some government agencies, particularly in high-income countries, and international bodies, there appears to be not only a general lack of awareness of the problems, but also no acceptance that there are problems. If the sustainability agenda is ever to succeed, especially in its aim to eliminate poverty, reducing human generated waste in general, and food and water waste in particular, needs to be taken seriously.

The world leaders, academic, and global communities—to some extent driven by social media influencers and mainstream media—appear to be manifestly concerned with climate change on a 24/7/365 basis, consistently and persistently evoking policies and legislation to combat the ill-effects of such changes while the waste problems are largely ignored, comparatively speaking. As previously mentioned, the 'Transforming Our World' document does not have prodigious 'waste' targets and statements, and over 90% of countries that have submitted their required nationally determined contributions (NDCs) to the Paris Agreement have not committed to reductions in food waste and loss [182]. This scenario is starting to change, likely prompted over the past decade by pivotal publications from the UN and academia reporting the significant contribution food waste and loss make to climate change [183, 184], and by more recent publications [185]. Eventually, in 2021, the Biden-Harris U.S. administration announced the launch of a "U.S. Methane Emissions Reduction Action Plan" which was quickly followed by a joint United States—EU "Global Methane Pledge" aimed at reducing methane emissions by at least 30% below 2020 levels by 2030 [186, 187].

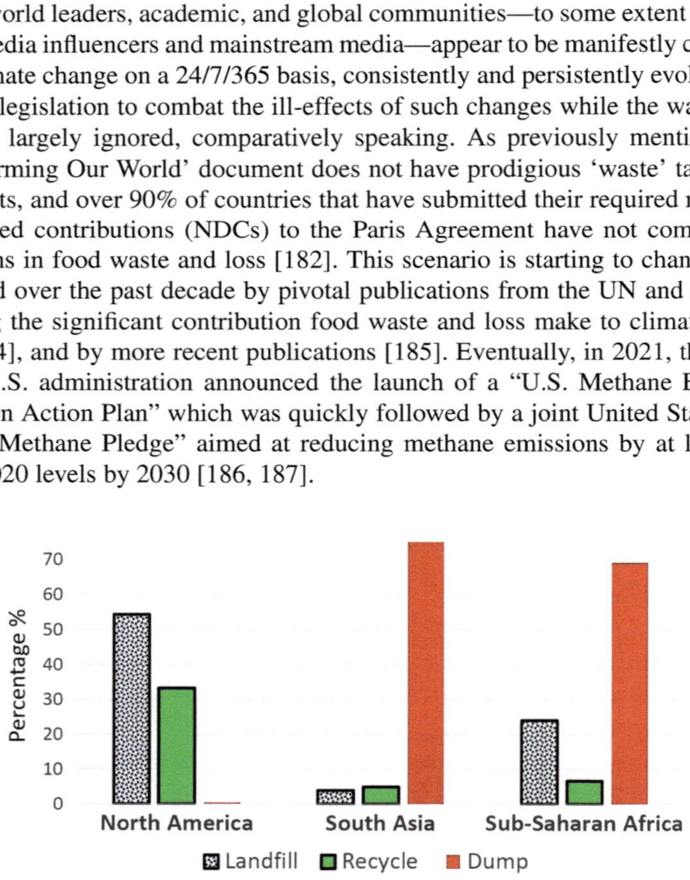

Fig. 1.14 MSW waste management comparisons; data from [16]

Over 150 countries have now signed on to this global pledge and although most of the associated media announcements identify the reduction of fossil fuel and energy sector emissions as the key part of the initiatives, agriculture is attracting more attention [188]. In some countries, India and Brazil are notable examples, the agricultural sector is the largest emitter of methane [187]. However, as was highlighted in the introductory remarks, the carbon footprint of food waste is a major contributor to climate change [13]. That being the case, is the reduction of food waste and its associated emissions a participant in the methane pledge? The answer is yes, since such emissions, particularly from so-called landfill gas (LFG), are a major contributor to anthropogenic methane emissions and, in some countries, LFG emissions are of a reasonably comparable scale to those of the energy and agriculture sectors and far greater than others such as coal-mining and 'manure management' [189]. It has to be remembered that GHG emissions are not solely carbon dioxide and that other constituents, such as methane, are usually expressed in carbon dioxide equivalents which can lead to the effects of the contributors being masked when only a global value of GHG is presented. In the case of organic waste (particularly food waste) deposited in landfills both carbon dioxide (CO_2) and methane (CH_4) are generated in roughly equal quantities and in a 'phased' manner, as discussed in the next section.

Determining the amount of GHG produced by MSW in general, and food waste in particular, from landfill sites is not an easy task depending upon the age, activity, and design of the facility, but it is far less of a challenge than attempting the measurement from both controlled and uncontrolled waste dump sites and, as shown in Fig. 1.14, in South Asia and Sub-Saharan Africa most of their waste is dumped. This likely means that the GHG generation is considerably higher than suggested by the known data acquired from sources in which, for instance, the SDG custodians have statistical confidence. Moreover, how can the GHG emissions from food losses be accurately measured? As food loss and food waste amounts are roughly comparable, on a global scale there could be a further underestimate of GHG generation particularly in those regions, illustrated in Fig. 1.13, where food losses are higher and coincidental with waste management relying largely on dumping. The causes from high regional proportions of food loss and waste being dumped were briefly discussed in Sect. 1.1.1. However, a noteworthy digital media detailing these dumping issues and the ramifications of the global economic and climate implications and food waste and loss, along with how one European country is tackling the challenges, was recently produced by students at Brown University in the United States and, in the author's opinion, provides a succinct introductory learning package for all interested or involved in 'waste-not: want not' [190].

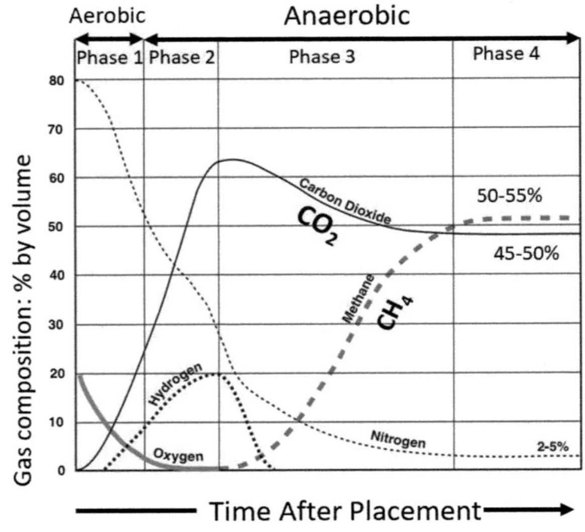

Fig. 1.15 The four temporal phases of LFG generation modified from [189, 191]

1.4 Food Waste and Climate Change

Prior to the GHG gases emitted from decomposing organic matter, including food waste, in landfill sites being predominantly associated with climate change, the concerns from the 1980s onwards, especially in the United States, were more focused on the possible environmental health issues of such emissions [191]. This led to the Agency for Toxic Substances and Disease (ATSDR), part of the United States Department of Health and Human Service (USHHS), publishing a 'Landfill Gas Primer' in 2001, dealing with a wide range of topics including the basics of what is LFG, monitoring and control, and the health and safety issues [191]. Most of the content is still relevant but has been updated by the USEPA and others, and is citied as a primary source in many publications dealing with various aspects of GHG generation from organic waste in landfill sites [187, 192, 193]. One of the most often cited items used from the primer[21] is a diagram illustrating the processes and timelines of the four phases of LFG production from bacterial action,[22] a slightly modified version of which is shown here as Fig. 1.15. It can be seen that CO_2 is produced from the waste almost immediately in the aerobic processes of 'phase 1' and continues into 'phase 4' of the aerobic processes, which could last over 20 years. Methane is not produced until all the oxygen in the waste has been consumed, which could take a year or more, and then increases rapidly eventually reaching a steady production rate slightly higher than that of CO_2. But what does all this mean?

[21] Chapter 2 of reference [191].

[22] Small amounts of other gases such as non-methane organic compounds can be generated in landfills by chemical and volatilization processes [189, 191].

If the LFG, as shown on Fig. 1.15, is not a single component gas and directly enters the atmosphere it will make a GHG contribution. However, CH_4 is a short-lived GHG but because its 'Global Warming Potential' (GWP[23])—a composite measure of a gas's potential to absorb energy and how long it remains in the atmosphere—is far higher than that of CO_2, it has a greater immediate impact on global warming, albeit, it quickly decays [12, 194, 195]. Nevertheless, exactly how CH_4 emissions impact the global surface temperatures, both in the near-term and longer-term, is a matter of discussion among the scientific community as there are concerns about the adequacies of the GWP approach in accessing its actual effects, which makes it difficult to directly attribute food waste emissions to precise increases in surface temperatures [196, 197]. While such issues have implications for the decision makers and consequent political policies, especially in formulating their Paris Agreement NDCs and the possible approaches to the Global Methane Pledge, from a waste reduction perspective the linkage between MSW, its food and other organic waste content, and climate change, can be said to have been globally acknowledged [198, 199].

Of course, the diagram showing the generation of LFG (Fig. 1.15) needs to be considered more as a generic representation rather than a depiction of the magnitude of the exact processes taking place in every landfill site as many variables are involved, such as the actual percentage of food and other organic waste contained in the MSW deposited at the site. The type and condition of such wastes at the time of deposit will also affect the rate of generation of the LFG. No doubt, the factors already discussed e.g., income, diet, size of the population being served, collection arrangements and so on, will also influence the characteristics of the GHG emissions in terms of the amounts and when they are released. Moreover, as shown in Fig. 1.14, not all MSW generated is disposed of in landfill sites, some will be recycled, composted, used to produce energy, or fed to undernourished people and animals as part of the waste hierarchy illustrated in Fig. 1.12. Over the last almost 60 years the United States have been measuring, gathering, and analysing MSW and landfill data and exemplars of this information are used in the following table, Table 1.7, to show the changing nature of landfilled MSW composition and, to some extent, such impacts as recycling and reuse [32].

Particularly noticeable is the increase in the amounts of plastics in the MSW streams which now account for the 2nd highest proportion of landfilled MSW. Overall, the percentage of annually generated MSW that is disposed of in landfills has sharply decreased from 94% in 1960 to 50% in 2018, but in terms of weight the amount has increased by almost 70%. However, perhaps the most concerning statistic is that, over the same time period, the amount of food waste sent to landfills annually is close to 3 times higher than in 1960 when the population was just over half of what it is today. At least for the United States, these data indicate that food waste is a serious issue and its contribution to global warming can hardly be ignored.

In these discussions of reducing the amount of MSW in general, and the amount of food waste and loss in particular, the impression can be gained that all the strategies and tactics are aimed at avoiding the use of landfills, albeit that landfills are preferable

[23] A relative numerical value used by climate science based on a datum of carbon dioxide = 1.

Table 1.7 United States exemplars of MSW generated and landfilled in 1960 compared to 2018 [32]

Materials[a]	Amount generated MSW[b] %	Amount landfilled[c] %
Paper and paperboard	23.1 (34.0)	11.8 (30.2)
Food	21.6 (13.8)	24.1 (14.8)
Plastics	12.2 (0.4)	18.5 (0.5)
Yard trimmings	12.1 (22.7)	7.2 (24.2)
Metals	8.8 (12.3)	9.5 (13.0)
ALL MSW	100	50 (94)

[a]The percentages given in parentheses are for 1960
[b]In 2018 292,360,000 US tons of MSW generated
[c]In 2018 146,120,000 US tons of MSW landfilled

to dumping of any kind. As illustrated in Table 1.7, the effects of recycling and other methods of using waste as a resource, rather than a burden, have been to reduce the percentage of MSW generation managed by landfills, but this is a relative measure since the actual weight of MSW diverted from, or directed to, landfills has increased substantially. Nevertheless, the concept of viewing waste as a resource has been gaining global momentum, particularly in Europe and in the Asia–Pacific regions, while in the Unites States the connections between natural resources, climate change, and waste are being firmly established [200–202]. In Europe, a detailed menu of directives and legislation have been published, in multiple languages, dealing with all aspects of waste reduction including its use as a resource, as perhaps a waypoint to the adoption of a circular economy [42, 203, 204]. However, it is inevitable that landfills will be required in the future just as they have been used in the past as a key contributor, sometimes the only contributor, to controlled waste management.

1.5 Landfills

In the ancient world, written records indicate that, generated waste was managed, long before the advent of landfills, usually by dumping or burying [205, 206]. There have even been suggestions that citizens of the Roman city of Pompeii, destroyed by volcanic action, didn't just dump their waste but collected it away from the city, then sorted and recycled it! [207]. These techniques continued, in the then developing world, for several millennia and while some communities provided specific locations for dumping, in general, waste was dumped where it was generated and scavenging was commonplace with the possibilities of spreading contamination and disease ever present. However, in the late nineteenth century some countries, e.g., the United States and the United Kingdom, started to incinerate waste and give local authorities legislated permission to collect waste. The purposeful burying of waste became more prevalent in the first-half of the twentieth century with local governments in particular having, in many instances, large holes dug in the ground into which the waste was dumped and eventually covered with soil. In essence a 'landfill' operation. The growing environmental scientific and public awareness of the second-half of the last century raised concerns not only about the unpleasant odours from open dump type landfill sites, but also their attraction to scavenging vermin and sometimes people. This led to governmental level actions in many countries in the form of legislation requiring that landfill siting and operation comply with certain legal and technical control procedures, such as the use of powered machinery to spread and compact the waste and cover it with soil daily. In many jurisdictions it was also forbidden to dump hazardous and industrial wastes and rubber products in these 'Controlled Landfills' which had to have a permit and permanent staffs to control the whole operation and ensure compliance [208].

Controlled landfills still exist today and their use by small and medium sized communities has proved to be an improvement of open dump sites [209]. However, such types of landfills were just the first step in improving the disposal and handling of waste, specifically MSW, because the potential and actual environmental problems associated with landfills became more apparent, such as the 'leachate' effluents, a collective term to describe the fluid percolating through the waste from various sources, including rainwater, with the potential to reach and contaminate groundwater resources which are used extensively in drinking water supplies and the surrounding soil quality [210]. These issues, coupled with the generation of LFG and its now acknowledged impact on climate change, have led to the need to upgrade the basic approach of controlled landfills to a more engineered design of landfills, not unexpectedly called 'Engineered Landfills' but frequently referred to as 'Sanitary Landfills' and in some cases 'Bioreactor Landfills' which are advanced and enhanced 'wet' versions of 'dry-tomb' sanitary landfills [208, 211, 212]. A sanitary landfill, in addition to the requirements of a controlled landfill, must have a way of collecting and treating leachate and the site must have 'hydrogeological' isolation, i.e., a liner which prevents leakage, particularly from the base of the landfill. Formal engineering designs have to be prepared and approved along with plans for how the waste is disposed of and how the site will be environmentally restored after the landfill's 'end-of-life' [213, 214]. Although, as with the controlled landfill, soil is used to cover the sanitary landfill, or a least part of it, during its operational life, now covers of polymeric materials are available which can be used in both the temporary and permanent coverage of engineered landfills [215].

A very simplified diagram of the basic structure of a modern landfill site is shown in Fig. 1.16 to illustrate the sanitary elements just discussed but a cross-section of an engineered landfill likely to be encountered in the United States is provided as Fig. 1.17 showing the level of engineering sophistication involved [216]. From an engineering viewpoint the adaptation of advanced landfill design and operation, especially with the use of bioreactors, has gained significant momentum over the past ten years or so, as ably described in recent review articles [217, 218]. Local authorities and the contracted constructors, albeit those with the capacity, technically and economically, have continuously improved the design, operation, and maintenance of new and updated MSW landfills. However, even in many high-income countries,

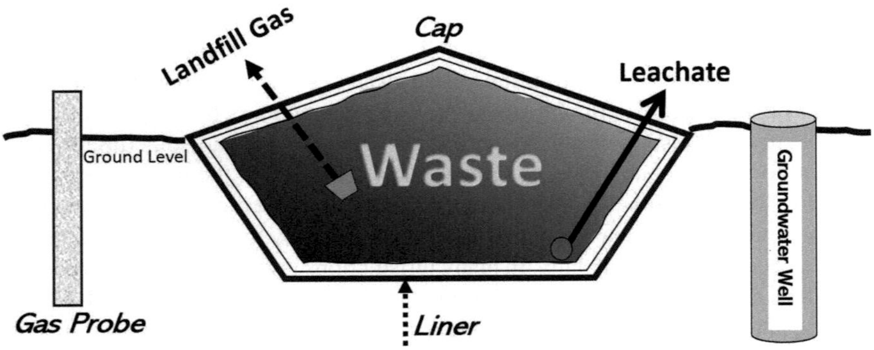

Fig. 1.16 Basic construction of a modern landfill site

as regulations become more stringent and the operating and initial capital costs of compliance increase, smaller communities struggle to afford the newer forms of landfill and the overall economic value of recycled materials is less than the costs of separation, collection, and treatment [219–221]. In the larger cities in the United States, especially since some Asian countries have started to reject U.S. 'waste' exports, an additional problem is the lack of land space for landfills, albeit the importing and re-exporting of discarded clothing provides significant economic benefits to other low-income countries [22, 222].

Engineered landfills are providing rich countries with an efficacious way of combating environmental health issues associated with MSW. Although they are the least preferred method of the food recovery hierarchy, Fig. 1.12, the rising costs of other elements of the recovery system, i.e., reduction, recycling, and reuse, likely means that more not less waste will be disposed of in such landfills in the future, at least in rich countries. In poorer countries as dumps are less expensive to establish and operate than landfills then, as shown, in regions such as Sub-Saharan Africa and South Asia, dumps are the dominant method of waste disposal, Fig. 1.14. This situation is driven largely by the lack of financial resources and millions of people live and work in and on waste dump sites with all the attendant ill-effects experienced in places like the United States in the early part of the last century [223, 224].

1.6 Concluding Remarks

In this chapter, a necessarily condensed overview of the use of the mantra 'waste not: want not', with a particular focus on food waste and MSW, has been discussed, with the impacts of the global disparities associated with national income levels, population size, and undernourishment highlighted. There appears to be no doubt that the relationships between the multitude of factors involved in the generation and management of waste are complex and at times puzzling, for example, why do

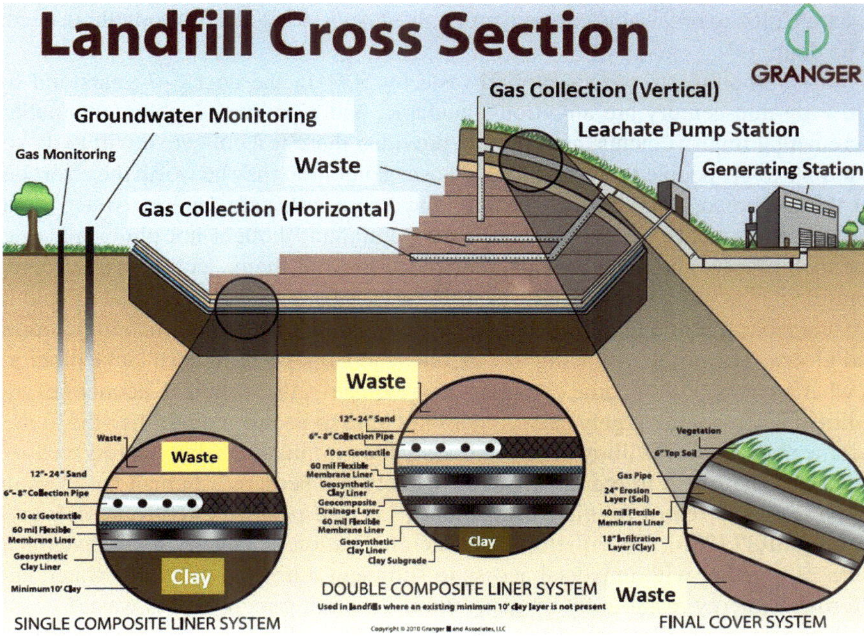

Fig. 1.17 Cross-section of a Granger[216] © Engineered Landfill Site with some font changes

high-income countries like Australia and Greece generate more per capita food waste than low-income countries like Ethiopia and India? Is it simply because they can? In terms of waste, the drive to achieve global sustainability, particularly the elimination of poverty, and simultaneously combat climate change aggressively are not always compatible. Yet, with the global prospects for ever increasing amounts of waste being generated, the negative influences on both sustainability and climate change goals means neither are likely to be accomplished. As humans, and human activities, are responsible for the generation of waste then why, apart from normal bodily effluents, should we not be accountable for significantly reducing the production of waste? Unhappily, there is both a general lack of awareness and of acceptance that waste generation, especially food waste, is a problem.

In attempting to identify and quantify the per capita generation of waste the cultural aspects and shared norms appear to regularly be ignored. For example, it is evident that the relative economic status of richer countries is not the only reason for their waste generation capacity as the extent is inherently influenced by their embrace of 'take-make-use-discard' and 'planned obsolescent' societal values. Such lifestyles are difficult to reconcile with the adoption of principles of sustainability, making a possible solution of a global migration to a circular economy, with the guiding principles as espoused by the Ellen MacArthur Foundation of 'eliminating waste, circulating waste and regenerating nature with the overall approach being buttressed by a transition to renewable energy', probably elusive [42, 204]. But something needs to be done.

The promptings to reduce global waste by 50% in the next 6–7 years and by 100% by mid-century are ambitious, laudable, and pipedreams, although smaller, incremental improvements, are possible provided there is a collective will to do so. Attempts to get people to 'rethink' their waste generation may be worth the effort but for those who are impoverished, hungry, have no access safe drinking water, clean cooking fuels or electricity, survival is the paramount thought not philosophizing. For such people, and there are hundreds of millions of them, acquiring engineered landfill sites is not a priority. However, in terms of waste reduction and its role in a sustainable future and climate change mitigation not everything has to be doom and gloom. The problems facing low income and LIFDCs have been tackled before in what are now high-income countries. Consequently, the technical, economic, and political wherewithal largely exists it does not have to be invented. If then the higher income countries are willing to help, not only their populations, but the less well-off regions of the world to reduce their waste problems perhaps a human based value analysis on the return on their investments, i.e., aid-priority effectiveness, would be insightful? [225]. Nevertheless, will the eradication of poverty and preventable diseases together with universal access to sufficient nutrition, drinkable water, and affordable energy result in future generations generating more or less waste?

If engineers of all persuasions, not just the civil, chemical, and environmental practitioners and educators, are to adapt to the waste generation and management issues, then the attitude of waste now; want not needs to be universally imbibed. While 'waste' education should preferably start in grade school, tertiary education can play an important role in providing a firm technical foundation on what can be done and what needs to be done. There is an encyclopedic knowledge bank on the subject of waste so there is no lack of learning materials. Hopefully, this chapter has supplied some insights into this topic and the importance of what could be described, arguably, as 'resource-use efficiency' in addressing the myriad of societal challenges caused by the somewhat uninhibited generation of waste.

Whatever actions are to be taken to reduce human generated waste, whether, domestic, industrial, or commercial, then exactly what constitutes waste needs to be defined, along with universally agreed methodologies for determining the amounts of the identified type of waste. Despite intense recent efforts to quantify and qualify waste, using SDG indicators, the data that is needed to accurately inform decision makers remains somewhat erratic.

References

1. United Nations, International Day For the Eradication of Poverty October 17, [Online]. Available: https://www.un.org/en/observances/day-for-eradicating-poverty. Accessed 2 May 2023

2. United Nations, *Observance of An International Day for the Eradication of Poverty; A/RES/ 47/196* (United Nations, New York, 1993)
3. UN, *Implementation of the Third United Nations Decade For the Eradication of Poverty (2028–2027); A/C.2/73/L.9* (United Nations, New York, 2018)
4. The Free Dictionary by Farlex, Waste Not, Want Not (2015). [Online]. Available: https://idioms.thefreedictionary.com/waste+not%2c+want+not. Accessed 2 May 2023
5. UNFCCC, *The Paris Agreement* (UNFCC, 2015)
6. United Nations, *The Sendai Framework for Disaster Risk Reduction 2015–2030* (United Nations, New York, 2015)
7. United Nations, *Transforming Our World: The 2030 Agenda for Sustainable Development* (United Nations, Washington D.C., 2015)
8. United Nations, *The Sustainable Development Goals Report 2022* (United Nations Publications, New York, 2022)
9. UNFCCC. Secretariat, *Nationally Determined Contributions Under the Paris Agreement: Synthesis Report by the Secretariat* (United Nations, New York, 2022)
10. International Science Council, *Report for the Mid-Term Review of the Sendai Framework for Disaster Risk Reduction* (International Science Council, Paris, France, 2023)
11. J. Hasell, M. Roser, E. Ortiz-Ospina, P. Arriagada, Poverty (2022). [Online]. Available: https://ourworldindata.org/poverty. Accessed 5 May 2023
12. H. Ritchie, How Much of Global Greenhouse Gas Emissions Come from Food? Our World in Data, 18 March 2021. [Online]. Available: https://ourworldindata.org/greenhouse-gas-emissions-food. Accessed 19 April 2023
13. UNFAO, *Food Wastage Footprint & Climate Change* (UNFAO, Rome, 2015)
14. GOV.UK, UK Statistics on Waste, 11 May 2022. [Online]. Available: https://www.gov.uk/government/statistics/uk-waste-data/uk-statistics-on-waste. Accessed 15 January 2023
15. USEPA, Learn the Basics of Hazardous Waste, 6 July 2022. [Online]. Available: https://www.epa.gov/hw/learn-basics-hazardous-waste. Accessed 26 April 2023
16. S. Kaza, L.C. Yao, P. Bhada-Tata, F. Van Woerden, *What a Waste 2.0: A Global Snapshot of Solid Waste Management to 2050. Urban Development* (World Bank Group, Washington, D.C., 2018)
17. M. Salinger, Climate variability and change: past, present and future—an overview. Clim. Change **70**, 9–29 (2005)
18. USEPA, Climate Change Indicators: Weather and Climate, 1 August 2022. [Online]. Available: https://www.epa.gov/climate-indicators/weather-climate. Accessed 27 April 2023
19. United Nations Office for Disaster Risk Reduction (UNDRR), *Global Assessment Report on Disaster Risk Reduction* (United Nations, Geneva, 2019)
20. J. Oetjen, V. Sundar, V. Sriram, K.R. Reicherter, A comprehensive review on structural tsunami countermeasures. Nat. Hazards **113**, 1419–1449 (2022)
21. vocabulary.com, Polysemous (2023). [Online]. Available: https://www.vocabulary.com/dictionary/polysemous. Accessed 1 May 2023
22. M. Bedat, Unravelled, *The Life and Death of a Garment* (Portfolio/Penguin, New York, New York, 2021), p. 312
23. USEPA, Reuse and Recycling Opportunities and Demolition, 7 April 2023. [Online]. Available: https://www.epa.gov/large-scale-residential-demolition/reuse-and-recycling-opportunities-and-demolition. Accessed 1 May 2023
24. M.T. Reporter, Brits have 55,000,000 unused mobile phones lying around, research finds, Associated Newspapers Limited, 26 November 2020. [Online]. Available: https://metro.co.uk/2020/11/26/brits-have-55000000-unused-mobile-phones-lying-around-research-says-13657334/. Accessed 1 May 2023
25. Wikipedia Contributers, Human waste, 18 April 2023. [Online]. Available: https://en.wikipedia.org/wiki/Human_waste. Accessed 3 May 2023
26. J. Kluger, How Poop Can Be Worth $9.5 Billion, 3 November 2015. [Online]. Available: https://time.com/4098127/human-waste-energy-recycling/. Accessed 5 May 2023

27. H. Ritchie, R. Max, Clean Water and Sanitation, OurWorldInData.org (2021). [Online]. Available: https://ourworldindata.org/clean-water-sanitation. Accessed 3 May 2023
28. World Bank Group, Sanitation, 6 October 2022. [Online]. Available: https://www.worldbank.org/en/topic/sanitation. Accessed 3 May 2023
29. A. Soliman, S.K. Jha, Closing the Access Gap for Water and Sanitation in Eastern and Southern Africa: Raising the Ambition, 27 March 2023. [Online]. Available: https://blogs.worldbank.org/water/closing-access-gap-water-and-sanitation-eastern-and-southern-africa-raising-ambition. Accessed 3 May 2023
30. World Bank Group, *Toward a Sustainable Energy Future for All: Directions for the World Bank Group's Energy Sector* (World Bank Group, Washington D.C., 2013)
31. USEPA, Wastes: What are the trends in wastes and their effects on human health and the environment?, 2 March 2023. [Online]. Available: https://www.epa.gov/report-environment/wastes#roe-indicators. Accessed 17 April 2023
32. USEPA, *Advancing Sustainable Materials Management: 2018 Tables and Figures: Assessing Trends in Materials Generation and Management in the United States* (USEPA, Washington, D.C., 2020)
33. UK Environmental Agency; SEPA; Cyfoeth Naturiol Cymru, *Waste Classification: Guidance on the Classification and Assessment of Waste* (1st Edition v1.2.GB); Technical Guidance WM3 (Bristol, Stirling & Cardiff, 2021)
34. Geneva Environmental Network, Achieving Zero Waste and the Role of Geneva, UNEP, 30 March 2023. [Online]. Available: https://www.genevaenvironmentnetwork.org/resources/updates/achieving-zero-waste-and-the-role-of-geneva/. Accessed 4 May 2023
35. K. Higgs, A Brief History of Consumer Culture, Curio, 11 January 2021. [Online]. Available: https://thereader.mitpress.mit.edu/a-brief-history-of-consumer-culture/. Accessed 4 May 2023
36. T. Wautelet, *Exploring the Role of Independent Retailers in the Circular Economy: A Case Study Approach* (Eufom European University for Economics & Management A.s.b.l., Luxembourg, 2018)
37. P.F. Kamba, M.E. Ireeta, S. Balikuna, B. Kaggwa, Threats posed by stockpiles of expired pharmaceuticals in low- and middle-income countries: a Ugandan perspective. Bull. World Health Organ. **95**(8), 594–598 (2017)
38. USFDA, Disposal of Unused Medicines: What You Should Know, 8 August 2022. [Online]. Available: https://www.fda.gov/drugs/safe-disposal-medicines/disposal-unused-medicines-what-you-should-know. Accessed 5 May 2023
39. G. Frascaroli, D. Reid, C. Hunter, J. Roberts, K. Helwig, J. Spencer, A. Escudero, Pharmaceuticals in wastewater treatment plants: a systematic review on the substances of greatest concern responsible for the development of antimicrobial resistance. Appl. Sci. **11**(15), 26 (2021)
40. M. Kostich, Pharmaceutical Residues in Municipal Wastewater, USEPA, 19 July 2022. [Online]. Available: https://www.epa.gov/water-research/pharmaceutical-residues-municipal-wastewater. Accessed 5 May 2023
41. L. Shearer, S. Pap, S.W. Gibb, Removal of pharmaceuticals from wastewater: a review of adsorptive approaches, modelling and mechanisms for metformin and macrolides. J. Environ. Chem. Eng. **10**(4), 13 (2022)
42. P. Özkan, Y.E. Karataş, Chapter 4; Linear Economy to Circular Economy: Planned Obsolescence to Cradle-to-Cradle Product Perspective, in *Handbook of Research on Entrepreneurship Development and Opportunities in Circular Economy* (Hershey/Beijing, IGI Global, 2020), pp. 61–86 (Chapter) Book 696
43. Automotive Recyclers Association, Get to Know Auto Recycling, Automotive Recyclers Association (2023). [Online]. Available: https://www.a-r-a.org/about-ara.html. Accessed 5 May 2023
44. Wikipedia contributors, Vehicle recycling, Wikipedia, The Free Encyclopedia, 11 April 2023. [Online]. Available: https://en.wikipedia.org/w/index.php?title=Vehicle_recycling&oldid=1149393376. Accessed 5 May 2023

45. C. Adams, Vehicle Recycling: A Driving Force in Sustainability, UCI, 20 May 2020. [Online]. Available: https://sustainability.uci.edu/vehicle-recyling-a-driving-force-in-sustainability/. Accessed 5 May 2023
46. M.Z.B.M. Saman, G.M. Blount, End of life vehicles recovery: process description, its impact and direction of research. Jurnal Mekanikal **21**, 40–52 (2006)
47. USEPA, Energy Recovery from the Combustion of Municipal Solid Waste (MSW), 9 February 2023. [Online]. Available: https://www.epa.gov/smm/energy-recovery-combustion-municipal-solid-waste-msw. Accessed 7 May 2023
48. International Energy Agency (IEA), Will energy from waste become the key form of bioenergy in Asia?, 10 January 2019. [Online]. Available: https://www.iea.org/articles/will-energy-from-waste-become-the-key-form-of-bioenergy-in-asia. Accessed 9 May 2023
49. P. Kaplan, J. Decarolis, S. Thorneloe, Is it better to burn or bury waste for clean electricity generation? Environ. Sci. Technol. **43**(6), 1711–1717 (2009)
50. M.J. Rogoff, F. Screve, *Waste-to-Energy: Technologies and Project Implementation*, 3rd edn (Elsevier, William Andrew, Norwich, New York, 2019), p. 228
51. S. Abbasi, The myth and the reality of energy recovery from municipal solid waste. Energ. Sustain. Soc. **8**(36), 15 (2018)
52. P. Hockenos, Waste to Energy—Controversial power generation by incineration, 26 May 2021. [Online]. Available: https://www.cleanenergywire.org/factsheets/waste-energy-controversial-power-generation-incineration. Accessed 7 May 2023
53. Canada Energy Regulator: Regie de l'energie du Canada, Provincial and Territorial Energy Profiles—Ontario, 3 March 2023. [Online]. Available: https://www.cer-rec.gc.ca/en/data-analysis/energy-markets/provincial-territorial-energy-profiles/provincial-territorial-energy-profiles-ontario.html. Accessed 9 May 2023
54. EnerData, China forecasts +6% in power consumption, 250 GW of new capacity in 2023, 24 January 2023. [Online]. Available: https://www.enerdata.net/publications/daily-energy-news/china-forecasts-6-power-consumption-250-gw-new-capacity-2023.html. Accessed 9 May 2023
55. K. Sieg, How Sweden Sends Just 1% of Its Trash to Landfills, 8 April 2022. [Online]. Available: https://reasonstobecheerful.world/waste-to-energy-sweden-power-plants/. Accessed 9 May 2023
56. SCB Statistics Sweden, Number of households in Sweden estimated with different methods, not given. [Online]. Available: https://www.scb.se/en/finding-statistics/statistics-by-subject-area/household-finances/income-and-income-distribution/households-finances/pong/tables-and-graphs/households/number-of-households/. Accessed 9 May 2023
57. B. Gardiner, In Europe, a Backlash Is Growing Over Incinerating Garbage, 1 April 2021. [Online]. Available: https://e360.yale.edu/features/in-europe-a-backlash-is-growing-over-incinerating-garbage. Accessed 9 May 2023
58. Deltaway, Waste-to-Energy: How-it-Works, [Online]. Available: https://deltawayenergy.com/2018/08/waste-to-energy-how-it-works/. Accessed 11 May 2023
59. A. Kanhar, S. Chen, F. Wang, Incineration fly ash and its treatment to possible utilization: a review. Energies **13**(24), 6681 (2020)
60. US Department of Transportation (USDoT), Fly Ash Facts for Highway Engineers (USDoT, Washington, D.C., 2003); Updated to 2017
61. T. Dale, What Is Fly Ash and How Is It Used in Concrete?, 4 March 2023. [Online]. Available: https://www.thespruce.com/fly-ash-applications-844761. Accessed 12 May 2023
62. L. Wang, Y. Jin, Y. Nie, R. Li, Recycling of municipal solid waste incineration fly ash for ordinary Portland cement production: a real-scale test. Resour. Conserv. Recycl. **54**(12), 1428–1435 (2010)
63. B.H. Cho, B.H. Nam, J. An, H. Youn, Municipal solid waste incineration (MSWI) ashes as construction materials—a review. Materials **13**(14), 30; Article #3143 (2020)
64. C. Marieta, A. Martín-Garin, Leon, A. Guerrero, Municipal solid waste incineration fly ash: from waste to cement manufacturing resource. Materials (Basel) **16**(6), 2538 (2023)

65. PACE-WEF, *A New Circular Vision for Electronics: Time for a Global Reboot* (WEF, Geneva, 2019)
66. C. Baldé, V. Forti, V. Gray, R. Kuehr, P. Stegmann, The Global E-waste Monitor 2017, United Nations University (UNU), International Telecommunication Union (ITU) & International Solid Waste Association (ISWA), Bonn/Geneva/Vienna (2017)
67. USEPA, Types of Composting and Understanding the Process, 3 April 2023. [Online]. Available: https://www.epa.gov/sustainable-management-food/types-composting-and-understanding-process. Accessed 15 May 2023
68. L. Smith, What to Know About How the Composting Process Works, 6 December 2022. [Online]. Available: https://www.webmd.com/balance/what-to-know-about-how-the-composting-process-works. Accessed 15 May 2023
69. A. Beck, How to Make Compost to Feed Your Plants and Reduce Waste, 1 May 2023. [Online]. Available: https://www.bhg.com/gardening/yard/compost/how-to-compost/. Accessed 15 May 2023
70. USEPA, Composting At Home, 22 November 2022. [Online]. Available: https://www.epa.gov/recycle/composting-home. Accessed 15 May 2023
71. G. Alexander, How Commercial Composting Works, 20 January 2022. [Online]. Available: https://earth911.com/business-policy/how-commercial-composting-works/. Accessed 15 May 2023
72. USEPA, *Composting Yard Trimmings and Municipal Solid Waste* (USEPA, Washington, D.C., 1994)
73. National Academies of Sciences, Engineering, and Medicine, *Achieving Highway Runoff Volume and Pollutant Reduction Using Vegetated Compost Blankets: A Guide* (The National Academies Press, Washington, D.C., 2023), p. 230
74. United States Department of Agriculture (USDA), Don't Waste Uneaten Food, [Online]. Available: https://www.usda.gov/sites/default/files/documents/usda-food-waste-infographic.pdf. Accessed 16 May 2023
75. World Food Programme, WFP at a glance, 2 March 2023. [Online]. Available: https://www.wfp.org/stories/wfp-glance. Accessed 16 May 2023
76. M. Roser, H. Ritchie, Hunger and Undernourishment Our World in Data (2019). [Online]. Available: https://ourworldindata.org/hunger-and-undernourishment. Accessed 17 April 2023
77. United Nations, *The Water-Energy-Food Nexus* (United Nations, Rome, 2014)
78. UNFAO, Water–Energy–Food Nexus (2023). [Online]. Available: https://www.fao.org/land-water/water/watergovernance/waterfoodenergynexus/en/. Accessed 16 May 2023
79. ESCAP and IISD, *Operationalizing the Environment-Health Nexus in Asia and the Pacific: A Policy Guide on Opportunities for Enhancing Health, Biodiversity, Food System and Climate Action* (UNESCAP/International Institute for Sustainable Development, Bangkok, 2022)
80. E. Haro, S.-M. Jarvensivu, J. Alilehto, P. Haravouri, *Electricity: How Long Could We Survive Without it?* (Sweco Group, Stockholm, 2019)
81. Lawrence Livermore National Laboratory, Energy Flow Charts: Charting the Complex Relationships among Energy, Water, and Carbon, March 2022. [Online]. Available: https://flowcharts.llnl.gov/sites/flowcharts/files/2022-09/Energy_2021_United-States.pdf. Accessed 17 May 2023
82. E. Commission, *2022 Report on the Achievement of the 2020 Energy Efficiency Targets* (European Commission, Brussels, 2022)
83. E. Parliament, *Energy Saving: EU Action to Reduce Energy* (European Parliament, Brussels, 2023)
84. European Union, Primary and Final Energy Consumption Slowly Decreasing, 21 January 2021. [Online]. Available: https://ec.europa.eu/eurostat/web/products-eurostat-news/-/DDN-20210128-1. Accessed 17 May 2023
85. Why is Energy Efficiency Important, 23 April 2023. [Online]. Available: https://www.climate-policy-watcher.org/energy-strategies/why-is-energy-efficiency-important.html. Accessed 17 May 2023

86. United Nations Department of Economic and Social Affairs, *World Population Prospects 2022: Summary of Results* (United Nations, New York, 2022)
87. Wikipedia contributors, Zero waste, 25 April 2023. [Online]. Available: https://en.wikipedia.org/w/index.php?title=Zero_waste&oldid=1151700195. Accessed 13 May 2023
88. N. Pietzsch, J.L.D. Ribeiro, J.F. de Medeiros, Benefits, challenges and critical factors of success for Zero Waste: a systematic literature review. Waste Manage. **67**, 324–352 (2017)
89. R. Abbe, ZWIA Joins the United Nations to Declare the International Day of Zero Waste—March 30, 2023, 23 March 2023. [Online]. Available: https://zwia.org/2023/03/zwia-joins-the-united-nations-to-declare-the-international-day-of-zero-waste-march-30-2023/. Accessed 18 May 2023
90. PST: Process Sensing Technologies, What Are the Key Measurements to Ensure Safety in Breathing Air, [Online]. Available: https://www.processsensing.com/en-us/blog/key-measurements-for-safety-in-breathing-air.htm. Accessed 22 May 2023
91. G.T. Reader, Clean air, clean water, clear conscience, in *Sustainable Engineering Technologies and Architectures* (AIP, Melville, New York, 2021), pp. 1–35
92. USEPA, Progress Cleaning the Air and Improving People's Health, 1 May 2023. [Online]. Available: https://www.epa.gov/clean-air-act-overview/progress-cleaning-air-and-improving-peoples-health. Accessed 23 May 2023
93. H. Ritchie, M. Roser, Air Pollution, January 2021. [Online]. Available: https://ourworldindata.org/air-pollution. Accessed 23 May 2023
94. World Health Organization, Air pollution data portal, 2023. [Online]. Available: https://www.who.int/data/gho/data/themes/air-pollution. Accessed 23 May 2023
95. USGS, Surface Runoff and the Water Cycle, 8 June 2019. [Online]. Available: https://www.usgs.gov/special-topics/water-science-school/science/surface-runoff-and-water-cycle. Accessed 23 May 2023
96. R. Stull, *Practical Meteorology: An Algebra-based Survey of Atmospheric Science*, Version 1.02b (University of British Columbia, Vancouver, B.C., 2017), p. 944
97. United Nations General Assembly, *64/292. The Human Right to Water and Sanitation* (United Nations, New York, 2010)
98. United Nations Department of Economic and Social Affairs, *Global indicator framework for the Sustainable Development Goals* (United Nations Statistics Division, New York, 2023)
99. UN Water, Roles and Responsibilities SDG Monitoring and Reporting, June 2017. [Online]. Available: https://www.unwater.org/news/roles-and-responsibilities-sdg-monitoring-and-reporting. Accessed 24 May 2023
100. UN Water, Data Compilation Process and Timeline by Indicator, 20 April 2023. [Online]. Available: https://www.unwater.org/sites/default/files/2023-04/Data%20compilation%20process%20and%20timeline_2023-04-20.pdf. Accessed 24 May 2023
101. Centers for Disease Control and Prevention (CDC), Assessing Access to Water & Sanitation, 24 March 2022. [Online]. Available: https://www.cdc.gov/healthywater/global/assessing.html. Accessed 24 May 2023
102. WHO, Drinking-Water, 21 March 2022. [Online]. Available: https://www.who.int/news-room/fact-sheets/detail/drinking-water. Accessed 25 May 2023
103. USEPA, *Best Practices to Consider When Evaluating Water Conservation and Efficiency As An Alternative for Water Supply Expansion* (USEPA, Washington, D.C., 2016)
104. USEPA, Using Water Efficiently: Ideas for Communities, USEPA, April 2000. [Online]. Available: https://www.epa.gov/sites/default/files/2017-03/documents/ws-ideas-for-communities.pdf. Accessed 25 May 2023
105. USEPA, Start Saving, 23 September 2022. [Online]. Available: https://www.epa.gov/watersense/start-saving. Accessed 25 May 2023
106. Water Footprint Calculator, Food's Big Water Footprint, https://gracecommunicationsfoundation.org/, 20 October 2022. [Online]. Available: https://www.watercalculator.org/footprint/foods-big-water-footprint/. Accessed 25 May 2023
107. M. Jalava, M. Kummu, M. Porkka, S. Siebert, O. Varis, Diet change—a solution to reduce water use? Environ. Res. Lett. **9**(7), 14 (2014)

108. A.-M. Codur, J.M. Harris, B. Roach, *Water: Economics and Policy* (Global Development and Environment Institute, Tufts University, Medford, MA, 2015), p. 47
109. S. Rehkamp, P. Canning, U.S. Food-Related Water Use Varies by Food Category, Supply Chain Stage, and Dietary Pattern, 9 August 2021. [Online]. Available: https://www.ers.usda.gov/amber-waves/2021/august/us-food-related-water-use-varies-by-food-category-supply-chain-stage-and-dietary-pattern/. Accessed 27 May 2023
110. T. Fox, *Global Food Waste Not, Want Not* (Institution of Mechanical Engineers, London, 2013)
111. Calories.info, Calories in Food: Calorie Chart Database, Calories.info, 2022. [Online]. Available: https://www.calories.info/. Accessed 27 May 2023
112. A.A. Olkowski, *Livestock Water Quality: A Field Guide for Cattle, Horses, Poultry and Swine* (Agriculture and Agri-Food Canada, Ottawa, 2009)
113. M. Giovanisci, The Beginner's Guide to Brewing Water Chemistry, 4 April 2022. [Online]. Available: https://www.brewcabin.com/brewing-water/. Accessed 29 May 2023
114. R.F. Stier, Assuring Water Quality and Safety in Food Processing, 1 August 2006. [Online]. Available: https://www.food-safety.com/articles/4624-assuring-water-quality-and-safety-in-food-processing. Accessed 29 May 2023
115. J. Mateo-Sagasta, S.M. Zadeh, H. Turral, More people, more food, worse water? a global review of water pollution from agriculture. The Food and Agriculture Organization of the United Nations and the International Water Management Institute, Rome/Columbo (2018)
116. H. Ritchie, M. Roser, Water Use and Stress, 2017. [Online]. Available: https://ourworldindata.org/water-use-stress. Accessed 30 May 2023
117. M. Stanley, A Deep Dive on the Water Crisis, 14 February 2022. [Online]. Available: https://www.morganstanley.com/ideas/water-scarcity-causes-and-solutions. Accessed 30 May 2023
118. Water Footprints Calculator, Water Use, Withdrawal and Consumption, 9 September 2022. [Online]. Available: https://www.watercalculator.org/footprint/water-use-withdrawal-consumption/. Accessed 30 May 2023
119. C. Dieter, M. Maupin, R. Caldwell, M. Harris, T. Ivahnenko, J. Lovelace, N. Barber, K. Linsey, Estimated use of water in the United States in 2015: U.S. Geological Survey Circular 1441 (U.S. Department of the Interior, U.S. Geological Survey, Washington, D.C., 2018)
120. A. Kohli, K. Frenken, *Cooling water for energy generation and its impact on national-level water statistics* (UNFAO-Aquastat, Rome, 2011)
121. USGS Science, Water-Use Terminology, 19 February 2019. [Online]. Available: https://www.usgs.gov/mission-areas/water-resources/science/water-use-terminology. Accessed 31 May 2023
122. A. Kohli, K. Frenken, C. Spottorno, *Disambiguation of water statistics* (UNFAO-Aquastat, Rome, 2012)
123. European Environmental Agency, Water abstraction by source and economic sector in Europe, 1 June 2022. [Online]. Available: https://www.eea.europa.eu/ims/water-abstraction-by-source-and. Accessed 1 June 2023
124. UNFAO, AQUASTAT—FAO's Global Information System on Water and Agriculture, 2021. [Online]. Available: https://www.fao.org/aquastat/en/overview/methodology/water-use. Accessed 1 June 2023
125. UNFAO, *The State of the World's Land and Water Resources for Food and Agriculture—Systems at Breaking Point* (UNFAO, Rome, 2022)
126. USDOA, *Determining Consumptive Use and Irrigation Water Requirements* (USDOA, Washington D.C., 1962)
127. WWAP (UNESCO World Water Assessment Programme), *The United Nations World Water Development Report Leaving No One Behind* (UNESCO, Paris, 2019)
128. K.H. Shaffer, D.L. Runkle, *Consumptive Water-Use Coefficients for the Great Lakes* (USGS, Reston, Virginia, 2007)
129. N. Nouri, F. Balali, A. Nasiri, H. Seifoddini, W. Otieno, Water withdrawal and consumption reduction for electrical energy. Appl. Energy **248**, 196–206 (2019)

130. UNDESA, *System of Environmental-Economic Accounting for Water* (United Nations, New York, 2012)
131. National Academies of Sciences, Engineering, and Medicine, *Science Breakthroughs to Advance Food and Agricultural Research by 2030* (The National Academies Press, Washington DC, 2019)
132. UNDESA, Ensure availability and sustainable management of water and sanitation for all, [Online]. Available: https://sdgs.un.org/goals/goal6. Accessed 6 June 2023
133. FAO and UN Water, *Progress on Change in Water-Use Efficiency. Global Status and Acceleration Needs for SDG Indicator 6.4.1* (UNWater, Rome, 2021)
134. UNFAO, FAO e-learning course on change in water-use efficiency (SDG 6.4.1), now available in multiple languages, FAO, 9 September 2022. [Online]. Available: https://www.fao.org/sustainable-development-goals/events/detail/en/c/1456597/. Accessed 7 June 2023
135. Law Insider, Consumptive Water Use Definition (2023). [Online]. Available: https://www.lawinsider.com/dictionary/consumptive-water-use. Accessed 7 June 2023
136. USEPA, Statistics and Facts, USEPA, 24 April 2023. [Online]. Available: https://www.epa.gov/watersense/statistics-and-facts. Accessed 7 June 2023
137. Gitnux, The Most Surprising Water Waste Statistics And Trends in 2023, 12 April 2023. [Online]. Available: https://blog.gitnux.com/water-waste-statistics/. Accessed 7 June 2023
138. P. Rozin, B. Haddad, C. Nemeroff, P. Slovic, Psychological aspects of the rejection of recycled water: contamination, purification and disgust. Judgm. Decis. Mak. **10**(1), 50–63 (2015)
139. S. Ricart, A.M. Rico, A. Ribas, Risk-Yuck factor nexus in reclaimed wastewater for irrigation: comparing farmers' attitudes and public perception. Water **11**(187), 20 (2019)
140. Smart Water Magazine, New surveys reveal that social acceptance of water reuse isn't the biggest challenge, 25 October 2021. [Online]. Available: https://smartwatermagazine.com/news/kwr/new-surveys-reveal-social-acceptance-water-reuse-isnt-biggest-challenge. Accessed 8 June 2023
141. UN, International Day of Awareness on Food Loss and Waste Reduction 29 September (2023). [Online]. Available: https://www.un.org/en/observances/end-food-waste-day. Accessed 21 May 2023
142. B. Lipinski, *SDG Target 12.3 On Food Loss and Waste: 2021 Progress Report* (World Resources Institute, Champions 12.3, Washington, D.C., 2021)
143. World Food Program USA, Food Waste vs. Food Loss: Know the Difference and Help #StopTheWaste Today (2023). [Online]. Available: https://www.wfpusa.org/articles/food-loss-vs-food-waste-primer/. Accessed 19 April 2023
144. UNFAO, *Voluntary Code of Conduct for Food Loss and Waste Reduction* (FAO, Rome, 2022)
145. UNFAO, Technical Platform on the Measurement and Reduction of Food Loss and Waste (2023). [Online]. Available: https://www.fao.org/platform-food-loss-waste/food-loss/food-loss-measurement/en. Accessed 21 May 2023
146. United Nations Environment Programme, *Food Waste Index Report 2021* (United Nations, Nairobi, 2021)
147. Worldometer, Countries in the World by Population (2023), 11 June 2023. [Online]. Available: https://www.worldometers.info/world-population/population-by-country/. Accessed 11 June 2023
148. D. Filipenco, World Waste: Statistics by Country and Short Facts, 7 March 2023. [Online]. Available: https://www.developmentaid.org/news-stream/post/158158/world-waste-statistics-by-country. Accessed 11 June 2023
149. SENSONEO, Global Waste Index 2022 (2023). [Online]. Available: https://sensoneo.com/global-waste-index/. Accessed 11 June 2023
150. World Bank Group, World Development Indicators: Size of the economy (2023). [Online]. Available: https://wdi.worldbank.org/table/WV.1. Accessed 11 June 2023
151. H. Ritchie, E. Mathieh, Interactive Charts on Waste Management (2023). [Online]. Available: https://ourworldindata.org/waste-management. Accessed 11 June 2023

152. N. Hamadeh, C. Van Rompaey, E. Metreau, S.G. Eapen, New World Bank Country Classifications by Income Level: 2022–2023, 1 July 2022. [Online]. Available: https://blogs.worldbank.org/opendata/new-world-bank-country-classifications-income-level-2022-2023. Accessed 10 June 2023
153. World Bank Group, Poverty and Inequality Platform, World Bank Group (2022). [Online]. Available: https://pip.worldbank.org/home. Accessed 12 June 2023
154. W. Kenton, International Poverty Line: Definition, Criticism, and Uses, 23 December 2020. [Online]. Available: https://www.investopedia.com/terms/i/international-poverty-line.asp. Accessed 12 June 2023
155. A. Peer, Global poverty: Facts, FAQs, and How to Help, 4 April 2023. [Online]. Available: https://www.worldvision.org/sponsorship-news-stories/global-poverty-facts#poverty-line. Accessed 13 June 2023
156. J. Hasell, From $1.90 to $2.15 A Day: The Updated International Poverty Line, 26 October 2022. [Online]. Available: https://ourworldindata.org/from-1-90-to-2-15-a-day-the-updated-international-poverty-line. Accessed 13 June 2023
157. U.S. Department of Health and Human Services, HHS Poverty Guidelines for 2023 (2022). [Online]. Available: https://aspe.hhs.gov/topics/poverty-economic-mobility/poverty-guidelines. Accessed 13 June 2023
158. H. Ritchie, M. Roser, Urbanization, September 2018. [Online]. Available: https://ourworldindata.org/urbanization. Accessed 13 June 2023
159. J. Vespa, L. Medina, D.M. Armstrong, Demographic Turning Points for the United States: Population Projections for 2020 to 2060 (U.S. Department of Commerce; Census Bureau, Suitland, MD, 2020)
160. Recycling Magazine, Waste Management and the Elderly: A Neglected Problem?, 3 July 2019. [Online]. Available: https://www.recycling-magazine.com/2019/07/03/waste-management-and-the-elderly-a-neglected-problem/. Accessed 14 June 2023
161. M. Gallant, Do Older Generations Waste Less Food?, 18 June 2018. [Online]. Available: https://guelphfoodwaste.com/2018/06/21/do-older-generations-waste-less-food/. Accessed 14 June 2023
162. Y.V. Fan, J.J. Klemeš, C.T. Lee, R.R. Tan, Demographic and socio-economic factors including sustainability related indexes in waste generation and recovery. Energy Sourc. Part A: Recov. Utilization Environ. Effects 14 (2021)
163. J. Parfitt, M. Barthel, S. Macnaughton, Food waste within food supply chains: quantification and potential for change to 2050. Phil. Trans. R. Soc. B **365**, 17 (3065–3081) (2010)
164. World Food Programme, 5 Facts About Food Waste and Hunger, 2 June 2020. [Online]. Available: https://www.wfp.org/stories/5-facts-about-food-waste-and-hunger. Accessed 15 June 2023
165. H. Ritchie, P. Rosado, M. Roser, Agricultural Production (2023). [Online]. Available: https://ourworldindata.org/agricultural-production. Accessed 15 June 2023
166. S. Ross, 4 Countries that Produce the Most Food, 30 April 2023. [Online]. Available: https://www.investopedia.com/articles/investing/100615/4-countries-produce-most-food.asp. Accessed 15 June 2023
167. UNDESA, Least Developed Countries (LDCs), United Nations, [Online]. Available: https://www.un.org/development/desa/dpad/least-developed-country-category.html. Accessed 16 June 2023
168. UNFAO, Hunger and Food Insecurity (2023). [Online]. Available: https://www.fao.org/hunger/en/. Accessed 18 June 2023
169. H. Ritchie, What is Undernourishment and How Is It Measured?, 4 February 2022. [Online]. Available: https://ourworldindata.org/undernourishment-definition. Accessed 16 June 2023
170. UNFAO, Low-Income Food-Deficit Countries (LIFDCs)—List updated June 2021 (2023). [Online]. Available: https://www.fao.org/countryprofiles/lifdc/en/. Accessed 18 June 2023
171. FAO, Ifad, UNICEF, WFP and WHO, ", The State of Food Security and Nutrition in the World, *Repurposing Food and Agricultural Policies to Make Healthy Diets More Affordable* (FAO, Rome, 2022), p.2022

172. J. Hasell, M. Roser, Famines, 7 December 2017. [Online]. Available: https://ourworldindata.org/famines. Accessed 18 June 2023
173. IPC, *The Integrated Food Security Phase Classification: What the IPC is* (IPC, Rome, 2022)
174. FEWS NET, Providing Evidence-Based Early Warning Information and Analysis of Food Insecurity and Its Drivers Worldwide [Online]. Available: https://fews.net/about. Accessed June 19 2023
175. FEWS NET, Monitoring & Forecasting Acute Food Insecurity (2023). [Online]. Available: https://fews.net/. Accessed 19 June 2023
176. UNEP, *Food Waste Index Report 2021: Appendix* (UNEP, Nairobi, 2021)
177. rts, *Food Waste in America in 2021: Statistics & Facts* (rts, 2021)
178. USEPA, United States Food Loss and Waste 2030 Champions, 21 May 2023. [Online]. Available: https://www.epa.gov/sustainable-management-food/united-states-food-loss-and-waste-2030-champions#list. Accessed 20 June 2023
179. USEPA, Food Recovery Hierarchy, 28 August 2022. [Online]. Available: https://www.epa.gov/sustainable-management-food/food-recovery-hierarchy. Accessed 20 June 2023
180. UNFAO, SDG Sub-Indicator 12.3.1.A—Food Loss Index (2020). [Online]. Available: https://elearning.fao.org/course/view.php?id=605. Accessed 19 April 2023
181. A. Murdock, What You Need to Know About Food Waste and Climate Change; A Surprisingly Large Problem with Some Simple solutions, 9 May 2017. [Online]. Available: https://www.universityofcalifornia.edu/news/what-you-need-know-about-food-waste-and-climate-change. Accessed June 21 2023
182. Y. Cabrera, Food Waste Needed in Nationally Determined Contributions, 14 November 2022. [Online]. Available: https://www.nrdc.org/bio/yvette-cabrera/food-waste-needed-nationally-determined-contributions. Accessed 21 June 2023
183. M.E. Pitesky, K.R. Stackhouse, F.M. Mitloehner, Chapter 1—Clearing the Air: Livestock's Contribution to Climate Change, in *Advances in Agronomy*, vol. 103, ed. by D.L. Sparks (Academic Press, Burlington, 2009), pp.1–40
184. UNFAO, *Global Food Losses and Food Waste—Extent, Causes and Prevention* (UNFAO, Rome, 2011)
185. J. Poore, T. Nemecek, Reducing food's environmental impacts through producers and consumers. Science **360**(6392), 997–982 (2018)
186. The White House, Fact Sheet: President Biden Tackles Methane Emissions, Spurs Innovations, and Supports Sustainable Agriculture to Build a Clean Energy Economy and Create Jobs, 2 November 2021. [Online]. Available: https://www.whitehouse.gov/briefing-room/statements-releases/2021/11/02/fact-sheet-president-biden-tackles-methane-emissions-spurs-innovations-and-supports-sustainable-agriculture-to-build-a-clean-energy-economy-and-create-jobs/. Accessed 24 June 2023
187. IEA, *Global Methane Tracker 2022* (IEA, Paris, 2022)
188. V. Volcovici, COP27: More Join Methane Pact as Focus Turns to Farms, 17 November 2022. [Online]. Available: https://www.reuters.com/business/cop/cop27-more-countries-join-methane-pact-focus-turns-farms-waste-2022-11-17/. Accessed 24 June 2023
189. USEPA, Basic Information about Landfill Gas, 21 April 2023. [Online]. Available: https://www.epa.gov/lmop/basic-information-about-landfill-gas. Accessed 24 June 2023
190. E.M. Dadourian, *The Global Food Waste Crisis and The Danish Solution* (Dadourian, Elise M, Copenhagen, 2020)
191. Agency for Toxic Substances and Disease Registry (ATSDR), Landfill Gas Primer: An Overview for Environmental Health Professional (United States Department of Health and Human Services, Buford, Atlanta, 2001)
192. USEPA, *LFG Energy Project Development Handbook* (USEPA, Washington D.C, 2021)
193. E. Ayandele, K. Huffman, M. Jungclaus, E. Tseng, D. Cusworth, R. Duren, B. Fisher, *Key Strategies for Mitigating Methane Emissions from Municipal Solid Waste* (RMI, Basalt, Colorado, 2022)
194. H. Ritchie, M. Roser, P. Rosado, CO_2 and Greenhouse Gas Emissions (2020). [Online]. Available: https://ourworldindata.org/greenhouse-gas-emissions. Accessed 25 June 2023

195. USEPA, Understanding Global Warming Potentials, 18 April 2023. [Online]. Available: https://www.epa.gov/ghgemissions/understanding-global-warming-potentials. Accessed 25 June 2023
196. M. Cain, Guest Post: A New Way to Assess 'Global Warming Potential' of Short-Lived Pollutants, 7 June 2018. [Online]. Available: https://www.carbonbrief.org/guest-post-a-new-way-to-assess-global-warming-potential-of-short-lived-pollutants/. Accessed 25 June 2023
197. A. Moseman, J. Trancik, Why Do We Compare Methane to Carbon Dioxide Over a 100-Year Timeframe? Are We Underrating the Importance of Methane Emissions?, 28 June 2021. [Online]. Available: https://climate.mit.edu/ask-mit/why-do-we-compare-methane-carbon-dioxide-over-100-year-timeframe-are-we-underrating. Accessed 25 June 2023
198. C.S. Malley, N. Borgford-Parnell, S. Haeussling, I.C. Howard, E.N. Lefèvre, J.C.I. Kuylenstierna, A roadmap to achieve the global methane pledge. Environ. Res.: Clim. **2**(1-011003), 15 (2023)
199. Environment and Climate Change Canada, *Reducing Methane Emissions from Canada's Municipal Solid Waste Landfills* (Environment and Climate Change Canada, Gatineau, QC, 2022)
200. European Environment Agency, Resource Efficiency and Waste, 17 June 2019. [Online]. Available: https://www.eea.europa.eu/themes/waste/intro. Accessed 26 June 2023
201. UNESCAP, *Valuing Waste, Transforming Cities* (United Nations, New York, 2015)
202. USEPA, Resources, Waste and Climate Change, 2 March 2023. [Online]. Available: https://www.epa.gov/smm/resources-waste-and-climate-change. Accessed 27 June 2023
203. EUR-lex, *Official Journal of the European Union, L 150* (Publications Office of the European Union, LUXEMBOURG, 2018)
204. Ellen MacArthur Foundation, What is a circular economy? [Online]. Available: https://ellenmacarthurfoundation.org/topics/circular-economy-introduction/overview. Accessed 21 April 2023
205. Commercial Zone Blog, A Brief History Of Waste Management, 9 July 2020. [Online]. Available: https://www.commercialzone.com/a-brief-history-of-waste-management/. Accessed 25 June 2023
206. F. Havlíček, M. Morcinek, Waste and pollution in the ancient roman empire. J. Landsc. Ecol. **9**(3), 33–49 (2016)
207. T. Machemer, Analysis of Pompeii's Garbage Suggests the Ancient Romans Recycled, Too, 27 April 2020. [Online]. Available: https://www.smithsonianmag.com/smart-news/pompeiis-trash-heaps-show-signs-recycling-system-180974755/. Accessed 27 June 2023
208. J. Jaramillo, *Guidelines for the Design, Construction and Operation of Manual Sanitary Landfills: A Solution for the Final Disposal of Municipal Solid Wastes in Small Communities* (Pan American Center for Sanitary Engineering and Environmental Sciences, Lima, Peru, 2003)
209. Open Learn Create, 10.2.4 Controlled Landfill, Open University (2017). [Online]. Available: https://www.open.edu/openlearncreate/mod/oucontent/view.php?id=80576&section=4.4. Accessed 27 June 2023
210. R. Nagarajan, S. Thirumalaisamy, E. Lakshumanan, Impact of leachate on groundwater pollution due to non-engineered municipal solid waste landfill sites of erode city, Tamil Nadu, India. Iran. J. Environ. Health Sci. Eng. **9**(1), 12 (2012)
211. R.H. Isenberg, D.D. Dillah, "Dry Tomb" Landfills—The Past, Present, and Possibilities, 9 February 2016. [Online]. Available: https://www.mswmanagement.com/home/article/13021893/dry-tomb-landfillsthe-past-present-and-possibilities. Accessed 27 June 2023
212. USEPA, Bioreactor Landfills, 6 November 2022. [Online]. Available: https://www.epa.gov/landfills/bioreactor-landfills. Accessed 27 June 2023
213. MIT.EDU, What is a Sanitary Landfill?, [Online]. Available: https://web.mit.edu/urbanupgrading/urbanenvironment/sectors/solid-waste-landfills.html. Accessed 27 June 2023
214. USEPA, Municipal Solid Waste Landfills, 4 April 2023. [Online]. Available: https://www.epa.gov/landfills/municipal-solid-waste-landfills. Accessed 27 June 2023

215. Industrial & Environmental Concepts, Inc, We Offer Custom Made Landfill Cover Products (2020). [Online]. Available: https://www.ieccovers.com/landfill-cover/. Accessed 27 June 2023
216. Granger Waste Services, Landfills are Designed to Protect Groundwater, Surface Water and Air Quality (2023). [Online]. Available: https://www.grangerwasteservices.com/wp-content/uploads/2016/02/cross-section_doublecompositeliner_16-1.pdf. Accessed 27 June 2023
217. G. Ozbay, M. Jones, M. Gadde, S. Isah, T. Attarwala, Design and operation of effective landfills with minimal effects on the environment and human health. J. Environ. Public Health **2021**, 13 (2021)
218. S. Nanda, F. Berruti, Municipal solid waste management and landfilling technologies: a review. Environ. Chem. Lett. **19**, 1433–1456 (2021)
219. L. Wade, I. Watch, Local Town Dumps a Thing of the Past Due to High Cost of Government Regulations, 17 December 2018. [Online]. Available: https://investigatemidwest.org/2018/12/17/local-town-dumps-a-thing-of-the-past-due-to-high-cost-of-government-regulations/. Accessed 28 June 2023
220. B. Arick, 3 Landfill Problems Costing your Business Money, 10 November 2022. [Online]. Available: https://www.roadrunnerwm.com/blog/3-landfill-problems-costing-your-business-money. Accessed 28 June 2023
221. Recycle Track Systems (RTS), Is Recycling Worth It? A Look At the Costs and Benefits of Recycling, 11 May 2021. [Online]. Available: https://www.rts.com/blog/is-recycling-worth-it/. Accessed 28 June 2023
222. R. Deer, Landfills: We're Running out of Space, 4 March 2021. [Online]. Available: https://www.roadrunnerwm.com/blog/landfills-were-running-out-of-space. Accessed 28 June 2023
223. International Samaritan, About Garbage Dump Communities (2023). [Online]. Available: https://www.intsam.org/about-garbage-dump-communities/. Accessed 28 June 2023
224. The Borgen Project Blog, Living in Landfills: Poverty in the Developing World, 19 July 2019. [Online]. Available: https://borgenproject.org/living-in-landfills-poverty-in-the-developing-world/. Accessed 28 June 2023
225. B. Lomborg, *Best Things First* (Copenhagen Consensus Center, Copenhagen, 2023), p.314

Chapter 2
Characterization and Management Practices of Solid Waste Generated from Hotels in Awka City, Nigeria

Obiora B. Ezeudu and Uzochukwu C. Ugochukwu

Abstract Steady growth in the tourism industry has led to generation of higher quantities of solid waste by hotels. Despite this fact, little has been studied about the characterization, composition, generation rates, and management methods that are adopted in tackling this solid waste in Nigeria. This work has set to provide an appropriate technical, institutional, financial, and legal, understanding of solid waste management systems by hotels in Awka, Nigeria. To achieve this aim, the current work applied a mixed method of in-depth, semi-structured interviews, field observation, and characterization analysis/quantification of solid waste. The results of the analyses show that the hotels generate solid waste which is exceptionally high in organic content and recyclables at 44.44% and 94.21% respectively. Other components include glass/bottles, textiles/leathers. WEEE, wood, rubber, sanitary, medical, polystyrene food pack, and metal were calculated at 5.22%, 2.80%, 1.18%, 1.10%, 1.99%, 0.60%, 0.11%, 8.60%, and 2.19% respectively. The generated waste quantities increase during the weekends. To improve the quality of solid waste management at the studied hotel, there is a great need for institutional rearrangement, policy improvements, and social awareness campaigns. There are also opportunities to be explored for achieving a circular economy.

2.1 Introduction

Although a disproportionate amount of waste is generated at the hotels, the scale and scope of related research remain limited [1]. This is particularly so in developing countries where the growth in the hotel business is ultimately associated with an

O. B. Ezeudu (✉)
Center for Environmental Management and Control, University of Nigeria, Enugu Campus, Enugu 410001, Nigeria
e-mail: obiezeudu@yahoo.com

U. C. Ugochukwu
Remediation and Environmental Risk Assessment Research Group, CEMAC, University of Nigeria, Enugu 410001, Nigeria

D. S-K. Ting and A. Vasel-Be-Hagh (eds.), *Engineering to Adapt*, Springer Proceedings in Energy, https://doi.org/10.1007/978-3-031-47237-4_2

increase in tourism, business, and recreational activities. Beyond providing lodging for tourists, hotels also serve as venues for both business meetings and recreational activities such as birthday parties, conferences, church programs, wedding ceremonies, seminars, workshops, etc. As such, a high number of visitors are expected in hotels daily, and consequently, the volume of waste generated is also expected to be high. In exact terms, it is estimated that 1 kg of solid waste is generated in a hotel room per night by a typical hotel guest [2, 3], while Earle and Townsend [4] calculated that solid waste generated in a hotel per day per room could range from 1.81 to 3.18 kg. The high volume of waste generated in hotels on daily basis is a precursor that solid waste management in hotels requires assiduous attention since the adverse impact of poorly managed waste on the hotel could be enormous. This could negatively affect the corporate image [5], organizational profit [6], market competitiveness, and customer loyalty [7].

There is limited comprehensive academic literature on solid waste management in hotels [1, 3, 8, 9]. But the majority of these studies are works conducted in developed countries with few case studies in developing countries. Only one work has studied solid waste management in hotels in the Nigerian city of Lagos [1]. Considering that Nigeria is one of the largest developing economies in Africa with a population estimated at 200 million people, the tourism industry plays a critical role in fostering economic growth as it is a highly strategic and profitable business sector. It is therefore important that more work needs to be conducted in other Nigeria locations towards building a solid conceptual and theoretical framework on solid waste management in the hotels in Nigeria and also to build a sustainable data bank that could aid in planning and policy-making. Contributing to this gap, the current work, therefore, studies the solid waste quantification, characterization, and waste management practices in the hotels in Awka, southeastern Nigeria. The following research questions were therefore raised to guide the research objectives:

(i) What are the quantification and characterization of solid waste streams generated in hotels in Awka?
(ii) What are the waste management practices adopted by hotels in Awka in managing their solid waste?
(iii) What are the institutional arrangement, financial mechanisms, and policy frameworks associated with solid waste management in hotels in Awka?

2.2 Materials and Methods

2.2.1 The Study Area

Awka is the capital city of Anambra state, located in southeast Nigeria. It is located 321.4 Km, from Abuja which is the federal capital of Nigeria while its geographical coordinates include latitudes 6.12° N and 6.21° N and between longitudes 7.06° E and 7.44° E. Awka is generally known to have two seasons, a dry season and a rainy season with moderate climate conditions. Awka has a population of 301, 657 according to the 2006 census figure, with a projected estimate of 2.5 million people as of 2018 [10]. Awka is close in proximity to Nnewi and Onitsha, two of the most commercialized cities in Nigeria. It has a federal university and the presence of several state and federal institutions that attract visitors from near and far. In exact terms, Anambra state is the third most urbanized state in Nigeria. A previous study has categorized Awka according to three zones [10]. The high-income zone (zone 1); middle-income zone (zone 2) and low-income (zone 3). Anambra State Waste Management Authority also divided the Awka metropolis into 17 catchment areas for ease of waste management. Setting out further criteria 17 hotels were selected from each of the catchment areas—6 from each zone 1 and 2, while 5 were from Zone 3.

2.2.2 *Study Method*

A mixed research method was adopted to achieve the aim of this study. Solid waste collected in hotels in Awka was sampled and characterized and quantified, then semi-structured interviews were conducted among respondents selected from key waste management stakeholders. The information on how the waste is managed (collected, sorted, transported, recycled, and/or disposed of) was extracted from the interviewed hotel managers to augment the waste characterization data for detailed analysis The choice of the respondents was guided by the need to present the views of the major stakeholders and/or individuals knowledgeable in solid waste management in the studied area [11].

2.2.3 *Waste Sampling Procedure*

We adopted ASTM D5231-92 Method, also known as the standard testing method for solid waste sampling [13, 14]. This method recommends that each unprocessed sorting waste sample should weigh between 91 and 136 kg. The number of samples required to be sorted was calculated using the formula:

$$N = \{ts/EX\}^2 \tag{2.1}$$

where t = student's t-test corresponding to the desired confidence level

$$t = (\overline{x} - \mu)/\left(s/\sqrt{n}\right) \tag{2.2}$$

S = Standard deviation

E = Desired precision level

X = Estimated Mean

The number of samples to be collected was statistically calculated in two stages. In the first stage, we use the student t-value of 1.645 at a 90% confidence level with E and S values of 10% and 0.03 respectively. The estimated mean, X at 0.10 corresponds to $n = \infty$ in the ASTM manual using organic food waste as the principal waste composition from the ASTM D5231-92. Then N was then determined as 24. In the second stage, the parameters were kept constant, the student t-value corresponding to $n = 24$ at 90% confidence level = 1.714 which gave a new n value of 26. Since 26 is within the 10% of 24, we adopted 26 as the number of samples required to be sorted at the three different stratified locations of Awka. The size of the waste container is an average of 50 to 130 kg. The waste was manually sorted to get the value of each component and the average taken to get the total waste at each dump site. Then the weights of the individual components were added to give the total quantity of waste at each dumpster/collection site. Samples were collected at least once a week for six months (from March 06 to 08 September 2022).

2.2.4 Waste Characterization and Quantification

There is no standardized method for waste collection and disposal in Awka municipality and there is no data or record for waste management planning. Based on this, the solid waste is sampled and characterized at the point of collection, at least once a week for six months at the designated locations. The weight of each respective sorted component was measured with the aid of weighing balance and the value was recorded. At the end of every sorting, the individual weights were summed together to arrive at the average daily total waste of the municipal solid waste at that location. If the collected waste stayed a day longer before quantification and characterization, the measured weight would be divided by the number of days it was dumped before sorting in that particular week. The percentage composition of each of the components was also calculated and documented. After determining the weight of each component, the collected waste was also separated according to recyclables and non-recyclables. The recyclability of each waste component was determined according to the following criteria: (i) recyclable waste that has an existing market such as metals, polythene bags, paper, and plastic (ii) those that can be recycled but has no existing market such as glass, leather, textile, and organic waste (iii) Those that cannot be recycled like medical waste, sanitary waste, and inert materials, etc. [13].

Table 2.1 Quantity of solid waste generated in hotels in Awka monthly (Kg/day)

Month	Average daily generation (Kg/day)
March	4980.60
April	5111.11
May	5081.21
June	5109.98
July	5123.77
August	6903.41
September	5554.31
Total Average	5409.19

2.3 Results and Discussion

2.3.1 *Waste Generation*

Results of the analysis show that the hotels in Awka generate about 5409.19 Kg of solid waste per day for 6 months studied. There were no significant variations in generations by months or seasons as shown in Table 2.1.

The values for August were high probably because there are usually more activities in southeast Nigeria during August. Variations in a generation were only noticed during the weekends. The waste quantities were observed to surge high during the weekends probably because activities are often at their peak in the hotels. Most of the hotels have their rooms filled at the weekends.

2.3.2 *Waste Characterization*

The composition by weight of waste generated in hotels in Awka is shown in Fig. 2.1. The highest constituent is organic which is 44.44%. This is followed by polythene and plastic which constitute 14.44 and 10.07 respectively. Paper, glass/bottle, metal, wood, rubber, textile/leather, Waste of Electrical and Electronics Equipment (WEEE), medicals, sanitary, and polystyrene food packs were calculated at 7.26%, 5.22%, 2.19%,1.10%, 1.99%, 2.80%, 1.18%, 0.11%, 0.60%, 8.60% respectively.

The composition of waste generated by the hotels in the three zones of Awka is shown in Fig. 2.2.

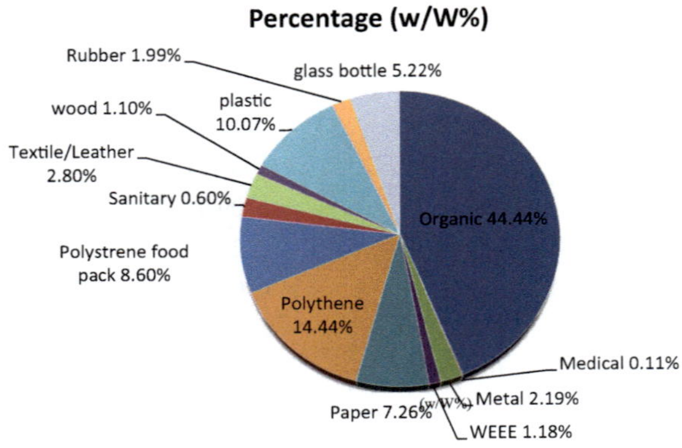

Fig. 2.1 The component analysis of the waste generated in hotels in Awka (w/W%)

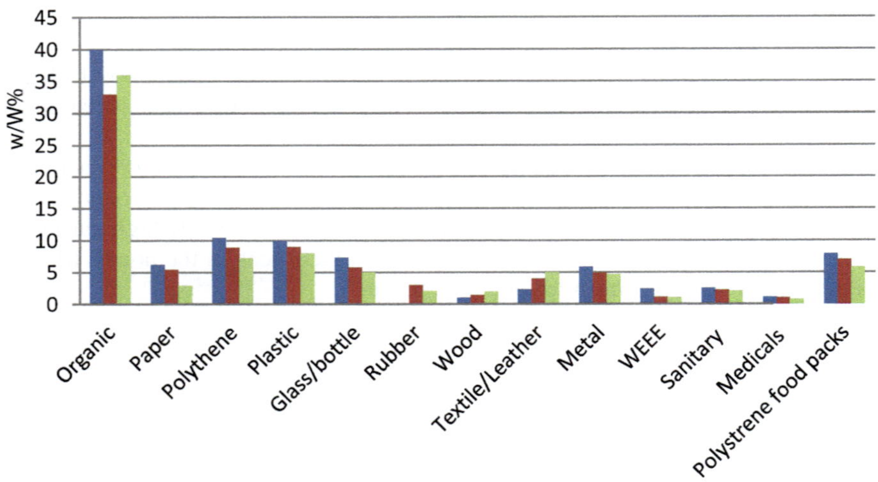

Fig. 2.2 Comparison of the waste generated at the three zones in Awka (w/W%)

2.4 Waste Management Practices in Hotels in Awka

2.4.1 Waste Collection

There are three main stages of waste collection practiced by hotels in Awka. According to one of the interviewed respondents

> "…first is the primary collection of waste by hotel cleaners within the hotel. The second is the collection of waste at the designated collection centers in the municipality. The third stage is the collection of the waste by the municipal authorities for onward movement to disposal sites".

In this arrangement, the hotel staff responsible for waste management will drop the generated waste from the hotel at the designated collection center for onward collection by the waste management authorities of the state. According to the interviewed hotel staff. "We package waste generated from the hotel on black polythene and drop them at the collection centers. …this waste collection is often conducted twice a week". There is no sorting or separation of waste from the point of generation. At the designated collection centers, informal waste pickers are often sighted picking waste with sellable values such as empty cans and bottles.

2.4.2 Waste Disposal and Treatment

The waste management authorities will collect the waste from the designated collection centers and transport them to dumpsites. There are no landfill facilities in Awka at the moment. Waste generated from different sources across the city such as households, businesses, markets, and institutions is collected as municipal solid waste by the waste management authorities and disposed of at the open dumpsites. Formal waste recycling schemes do not exist in Nigeria, and there are no available novel waste treatment facilities in Awka for the sustainable management of waste in the state [12].

2.4.3 Recommendation and Conclusion

Quantification, characterization, and waste management practices in the hotels in Awka as analyzed have shown that an enormous amount of organic waste and recyclables are generated. This is an opportunity for the responsible authorities to establish modalities for resource recovery/recycling initiatives from these wastes for the creation of a circular economy. A waste composting facility that can convert the organic component to compost is recommended. This can be achieved through policy improvement, institutional reorganizations, and awareness campaigns.

References

1. V. Filimonau, C.O. Tochukwu, Exploring managerial approaches to mitigating solid waste in hotels of Lagos, Nigeria. J. Clean. Prod. **270**, 122410 (2020)
2. P. Bohdanowicz, Environmental awareness and initiatives in the Swedish and Polish hotel industries: survey results. Int. J. Hosp. Manag. **25**, 662–682 (2006)
3. M. Abdulredha, R.A. Khaddar, D. Jordan, P. Kot, A. Abdulridha, Estimating solid waste generation by hospitality industry during major festivals: a quantification model based on multiple regression. Waste Manage. **77**, 388–400 (2018)

4. J.F.K. Earle, J.M. Townsend, Florida's Pilot Hotel/Motel Recycling Project: Final Report (1991). Available online: https://scholar.google.com/scholar?cluster=11226656645650769295&hl=en&as_sdt=2005&sciodt=0,5
5. V. Fillimonau, D.A de Coteau, Food waste management in hospitality operation: a critical review. Tour. Manage. **71**, 234–245 (2019)
6. E. Papargyropoulos, J. Wright, R. Lozano, K. Steinberger, R. Padfield, N. Ujang, Conceptual framework for the study of food waste generation and prevention in the hospitality sector. Waste Manage. **49**, 326–336 (2016)
7. S.I. Pirani, H.A. Arafat, Reduction of food waste generation in the hospitality industry. J. Clean. Prod. **132**, 129–145 (2016)
8. W. Chaabane, A. Nassour, M. Nelles, Solid waste management key indicator development for hotels: a Tunisian case study analysis. Recycling **3**, 56 (2018)
9. H.R. Radwan, E. Jones, D. Minoli, Managing solid waste in small hotels. J. Sustain. Tour. **18**, 175–190 (2010)
10. O.B. Ezeudu, C.G. Ozoegwu, C.N. Madu, A statistical regression method for characterization of household waste: a case study of Awka municipality in Nigeria. Recycling. **4**, 1 (2019)
11. T.C. Nzeadibe, Solid waste reforms and informal recycling in Enugu urban areas, Nigeria. Habitat Int. **33**, 93–99 (2009)
12. O.B. Ezeudu, J.C. Agunwamba, U.C. Ugochukwu, T.S. Ezeudu, Temporal assessment of municipal solid waste management in Nigeria: prospects for circular economy adoption. Rev. Environ. Health **36**, 327–344 (2021)
13. C.O. Ugwu, C.G. Ozoegwu, P.A. Ozor, Solid waste quantification and characterization in university of Nigeria, Nsukka Campus, and recommendations for sustainable management. Heliyon **6**, e04255 (2020)
14. ASTM D5231, Standard testing method for determination of the composition of unprocessed municipal solid waste. Available: https://www.astm.org/d5231-92r16.html. Assessed 20 Feb 2023

Chapter 3
Ignition and Combustion Characteristics of High-Pressure DME Spray Under Diluted Conditions

Binghao Cong, Simon LeBlanc, Xiao Yu, and Ming Zheng

Abstract As a renewable energy source, dimethyl ether presents a multitude of advantages as a compression ignition fuel. The upcoming emissions regulations will further challenge the internal combustion engine to realize emission reduction at the tailpipe. The use of exhaust gas recirculation has proven an effective strategy to lower NOx emissions via weakened charge reactivity and lowered bulk combustion temperatures. This is especially beneficial in a heavy-duty engine powered by diesel fuel that is susceptible to a NOx-soot trade-off because of the highly heterogeneous combustion characteristic. However, additional weakening of the cylinder charge is likely required to further constrain NOx formation. As a result, dimethyl ether has gained attraction for its beneficial properties as an alternative to diesel fuel, namely high reactivity, volatility, and fuel-borne oxygen content. Therefore, interest surrounds the ignitability and combustion stability of dimethyl ether combustion at dilution oxygen concentrations. In this work, two empirical setups were used to study the ignition and combustion behaviour of high-pressure dimethyl ether fuel injection. A constant volume combustion chamber was employed to study the ignition behaviour under different background temperatures using high-speed video recordings. A single-cylinder engine platform was employed to study the combustion behaviour under different background temperatures and diluted conditions. Diesel fuel was tested under similar testing conditions as the baseline for comparison. The imaging results indicated that dimethyl ether presented a more distributed burn development and higher resilience to harsh background environments. The results from engine studies showed that an improvement in engine stability was achieved by increasing the fuel injection pressure. Moreover, dimethyl ether realized slightly improved engine stability than diesel under highly diluted conditions.

B. Cong (✉) · S. LeBlanc · X. Yu · M. Zheng
Department of Mechanical, Automotive, and Materials Engineering, University of Windsor, 401 Sunset Ave, Windsor, ON N9B 3P4, Canada
e-mail: congb@uwindsor.ca

D. S-K. Ting and A. Vasel-Be-Hagh (eds.), *Engineering to Adapt*, Springer Proceedings in Energy, https://doi.org/10.1007/978-3-031-47237-4_3

3.1 Introduction

Compression ignition (CI) engines dominate the heavy-duty transportation sector for their high efficiency and power density. Diesel fuel is the principal energy source for such systems for its superior energy density, convenient handling, and abundance of feedstock. However, government regulations and decarbonization concerns draw more attention for internal combustion engine manufacturers in the past decades. For heavy-duty diesel engines, nitrogen oxides (NOx) and particulate matter (PM) are especially challenging to meet the upcoming emission regulations. NOx formation rates are strongly dependent on temperature and therefore reducing combustion temperatures through exhaust gas recirculation (EGR) has become an effective tool to meet regulations [1]. The recirculated exhaust gas (mainly CO_2 and water) diluted the intake charge and lowers the oxygen concentration, which prolongs the ignition delay, leading to more local regions of homogeneous mixing quality [2]. In turn, the formation of high-temperature pockets is reduced. However, diesel combustion is restricted to moderate EGR rates as the PM emissions rise, often referred to as the NOx-soot trade-off [2, 3]. Under high EGR rates, low-temperature combustion can be achieved to realize low NOx and PM emissions, simultaneously, but the combustion efficiency and stability deteriorate drastically [4].

Dimethyl ether (DME) has attractive properties which make it plausible to tackle the challenges faced with low-emissions diesel combustion. Primarily, DME is a highly reactive fuel whose combustion produces near-zero PM, owing to high oxygen content (34.8%), high volatility, and a simple chemical structure [5, 6]. Park et al. [6] investigated the performance and emission characteristics of the DI compression ignition (CI) engine which operated on DME using intake boost and EGR dilution. The power outputs for DME and diesel operated engines are similar at matching test conditions. Sidhu et al. [7] performed a deep analysis of the particulate formation from the combustion of oxygenated and non-oxygenated fuels in a premixed gas phase and a spraying phase. Notably, DME showed >95% fewer PM and its smoke filter paper appeared brown rather than black particles from diesel. Sato et al. [8, 9] investigated the performance and emission characteristics of DME on a DI diesel engine. With the volatility and ultra-low smoke tendencies of DME, higher EGR rates were allotted and realized up to a 2/3rd reduction in NOx emissions [8]. In addition, the combustion efficiency showed significant improvements in comparison with a diesel under low loads [9].

The unique fuel properties of DME summarized in Table 3.1 suggest that combustion will be efficient and clean. Supporting empirical studies of the fuel spray behaviour of DME under inert and reactive conditions progress the development of an optimized DME engine. Teng et al. [10] note that the ignition delay period comprises a physical and chemical process, wherein the physical delay dominates the total ignition delay period. The findings showed the evaporation of DME droplets to be 2~3 times faster than diesel, assuming a similar droplet diameter. The ignition delay period was shorter and the cetane number was estimated at 68, albeit literature often states a cetane number between 55 and 68 [6, 10–12]. Mitsugi et al. [13]

Table 3.1 Select fuel properties of DME and diesel

Property	DME	Diesel
Chemical formula	C_2H_6O	$C_nH_{1.87n}$
Cetane number	55–68	45–50
Energy density (LHV)	28.4 MJ/kg	42.5 MJ/kg
Oxygen content (by mass)	34.8%	0
Auto-ignition temperature	453 K	483 K
Vapour pressure	530 kPa	<<10 kPa
Liquid density @ 15 °C	667 kg/m^3	831 kg/m^3

observed inert (43 bar and 444 K) and reactive (60 bar and 920 K) DME fuel sprays in a constant volume combustion vessel under. The DME spray showed rapid evaporation and a non-luminous flame corresponding with the ultra-low sooting tendencies of DME combustion. The usage of DME as engine fuels also introduce some challenges to the high-pressure injection systems. The DME fuel has a much lower viscosity than diesel, meaning that DME can cause wear problems to the injectors. Therefore, an advancement in injection hardware is necessary prior to widespread usage in heavy-duty engines.

The benefits of DME combustion are apparent in literature works. Conversely, the combustion behaviour of DME fuel as compared with diesel fuel is limited. A direct comparison between DME and diesel fuel may present details of the challenges or advantages that may arise in the control or calibration of DME engines. In this work, a fundamental comparison of the ignition and combustion characteristics between DME and diesel fuel was conducted. The ignition characteristics were studied by video recording of a high-pressure DME reactive fuel spray into an oxygen-enriched and elevated temperature background. Thereafter, combustion characteristics were analyzed with a single-cylinder engine adapted for high-pressure DME operation. In all studies, matching conditions for diesel fuel were compared.

3.2 Experimental Setup

The ignition characteristics and combustion behaviour of DME were investigated using a constant volume combustion chamber and single-cylinder engine, respectively. The former research platform will present visual data into the ignition process of high-pressure DME flame using high-speed video recording in a constant volume combustion chamber. The latter research platform will analyze the combustion behaviour using high-pressure direct injection DME in a single-cylinder research engine.

3.2.1 Constant Volume Chamber Platform

Reactive fuel spray processes were captured in an optically accessible constant volume combustion chamber, as shown in Fig. 3.1. The chamber was machined from a block of stainless steel with slots for various adapters such as windows, injectors, heater cores, spark plugs, and sensors. The LabVIEW real-time (RT) hardware and software environment were used to monitor the system parameters, such as chamber pressure; deliver control signals such as ignition and injection events, and acquire data such as chamber pressure history.

A high-pressure and high-temperature fuel spray background environment was realized with an ignitable mixture. The ignitable mixture was supplied by a premixing reservoir with a user-controlled distribution of ethylene-oxygen–nitrogen, as shown in Table 3.2. The combustion reactants and products determine the background species that the fuel interacts with during spray dispersion. In these studies, the ethylene reactants concentration and oxygen products concentration were fixed at 5% and 15%, respectively. The selected oxygen concentration represents a medium EGR diluted conditions. For all tests, the chamber was maintained at 80 °C to limit window water condensation after ethylene combustion.

At the time of spark ignition, the chamber pressure rose rapidly during ethylene combustion (pre-burn period) followed by a steady drop due to heat transfer (post-burn period), as shown in Fig. 3.2. The background temperature was approximated as the bulk gas temperature using the ideal gas law with constant density. A high-pressure fueling system with a pneumatic pump connected to a diesel rail was configured to the chamber. The power electronics necessary for piezoelectric fuel injection were controlled by an injection driver (EFS IPoD) and triggered by the

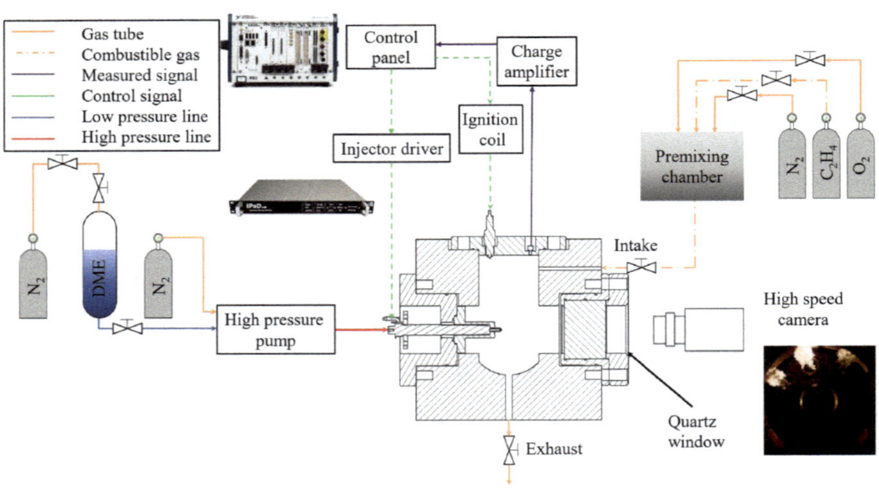

Fig. 3.1 Constant volume combustion chamber system

Table 3.2 Chamber conditions at the spark ignition (pre-burn stage) and fuel injection (post-burn stage) events

		Spark ignition	Fuel Injection
		Pre-burn	Post-burn
Ambient composition			
Ethylene (C_2H_4)	Mole fraction	5	0
Nitrogen (N_2)	Mole fraction	65	65
Oxygen (O_2)	Mole fraction	30	15
Water (H_2O)	Mole fraction	0	10
Carbon dioxide (CO_2)	Mole fraction	0	10
Ambient conditions			
Temperature	K	353	1100–1500
Pressure	bar	8	22–35
Density	kg/m^3	7.56	7.56

control system. The injection signal was synchronized to the camera for precisely timed video recording. A background temperature of 900 K up to 1800 K could be realized by adjusting the injection timing appropriately.

To prepare for each fuel spray test, the chamber was vacuumed and then supplied with the ignitable mixture (refer Table 3.2) up to 8 bar. Following the pre-burn stage, an injection event is triggered at a background temperature of 1237 K, 1168 K and 1125 K, which aligned with a timing of 1000 ms, 1200 ms, and 1300 ms, respectively. The recorded videos were analyzed and processed for improved contrast for visual comparison amongst conditions. The approaches to imaging analysis included plume detection, isolation, image cropping, and a false-colour representation. A single plume was focused on the measurements and the same plume was used throughout the

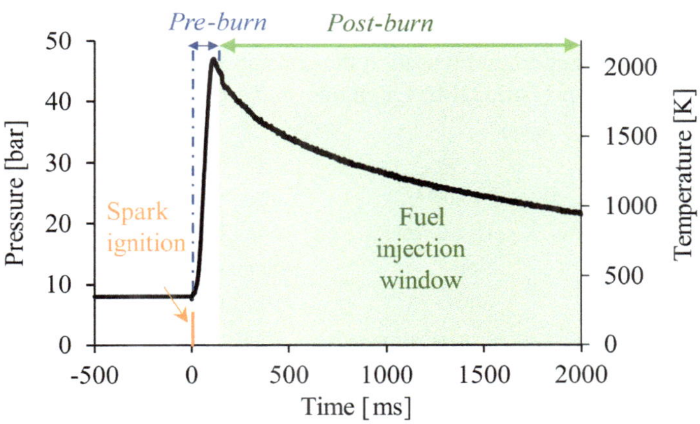

Fig. 3.2 The timeline of chamber pressure for the reactive fuel spray studies

analysis. Primarily, the ignition delay and lift-off length were quantitatively compared as both can provide valuable insights into ignition characteristics. Ignition delay and the corresponding lift-off length at the onset of autoignition were measured to quantify the ignition characteristics of both fuels. The ignition delay was defined as the time between the start of injection and the first appearance of autoignition, i.e., flame chemiluminescence. DME and diesel fuel were subject to matching fuel conditions for direct comparison, fixed 450 bar injection pressure and 600 μs injection duration.

3.2.2 Single-Cylinder Engine Platform

The combustion experiments were conducted using a CI direct injection (DI) single-cylinder research engine, with specifications summarized in Table 3.3. The engine crankshaft was coupled with a direct current dynamometer to control the engine speed. The intake air was supplied by a dry and oil-free air compressor and adjusted by an electro-pneumatic regulator for the intake pressure control. Two large air reservoirs in the intake and exhaust path were installed to minimize the pressure fluctuations during the 1200 rpm engine operation. A volumetric flow meter paired with pressure and temperature sensors was used to measure the intake mass air flow rate. To minimize the discrepancies in the testing results, the engine coolant temperature was maintained at 80 °C, and the oil temperature and pressure are maintained at 80 °C and 250 kPa, respectively.

The in-cylinder pressure was measured using a flush-mounted piezoelectric dynamic pressure transducer. The crankshaft position was monitored using a rotary encoder mounted to the crankshaft. The cylinder pressure acquisition and engine control system were configured using LabVIEW hardware and software environment. Real-time controllers embedded with FPGA devices were used to link the sensors and actuators, allowing for prompt, deterministic, and parallel control. Therefore, the pressure history measurement was resolved at 0.1°CA (7200/cycle). Each representative data point was an average of 200 consecutive cycles.

The same fuel injection pump used in the constant volume chamber was adapted as an off-board high-pressure DME DI engine study. Further details can be found in

Table 3.3 Engine specifications

Engine type	Single-cylinder 4-stroke
Displacement	0.744 L
Bore	95 mm
Stroke	105 mm
Compression ratio	16.5:1
Piston bowl type	Deep omega
Injector	Piezoelectric 7-hole with ∅ 0.200 mm
Engine speed	1200 rpm

previous work [12, 14]. The single-plunger injection pump shows significant pressure variations during injections at a greater frequency than >1Hz. Therefore, a large reservoir (300 mL) was installed between the injection pump and diesel rail to increase the high-pressure fuel reserve and minimize pressure fluctuations. With this system, continuous operation at 1200 rpm engine speed, equivalent to 10 Hz injection frequency, can realize a lower than 1% coefficient of variation in the engine load (IMEP).

The conditions for this study were achieved with exhaust gas recirculation (EGR), wherein the exhaust gases (after the catalyst) were cooled and recirculated into the intake manifold. The EGR rate was controlled by a back pressure control with a fixed EGR valve position. The EGR rate was quantified by the ratio reduction in fresh air intake charge. Combustion metrics, such as the indicated mean effective pressure (IMEP), apparent heat release rate (HRR), and the crank angle of 50% heat release (CA50), were computed based on the measured data. For efficiency metrics, the exhaust gases (before the catalyst) were measured with an infrared spectrometer (FTIR). A smoke meter was used to measure the soot in the exhaust.

3.3 Results and Discussion

The qualitative results ignition characteristics investigation in the constant volume chamber and quantitively combustion behaviour studies in the single-cylinder engine platform are shared and discussed in this section. The ignition characteristics for diesel and DME fuels are compared with a series of images captured by the high-speed cameras. The combustion behaviour is analyzed with the combustion performance parameters (apparent heat release rate, combustion phasing) and measured engine-out emissions. The recorded images were processed for improved comparison among test points, as shown in Fig. 3.3. The approaches for imaging analysis included plume isolation, image cropping, and a false-color representation was applied to increase contrast in the flame intensities. The light intensities were mapped into color-coded. This work focused on the same plume among each test case (circled) because the premixed combustion created a non-uniform temperature distribution within the chamber.

3.3.1 Ignition Characteristics

Figure 3.4 shows a temporal sequence of the DME fuel spray and ignition processes at an ambient temperature of 1237 K. It is important to mention that the injection signal was commanded at 0 ms, while the injected liquid fuel was first observed at 0.37 ms. The interval between the start of the injection signal and the actual start of injection (nozzle lift) is referred to as the injector opening delay, which represents

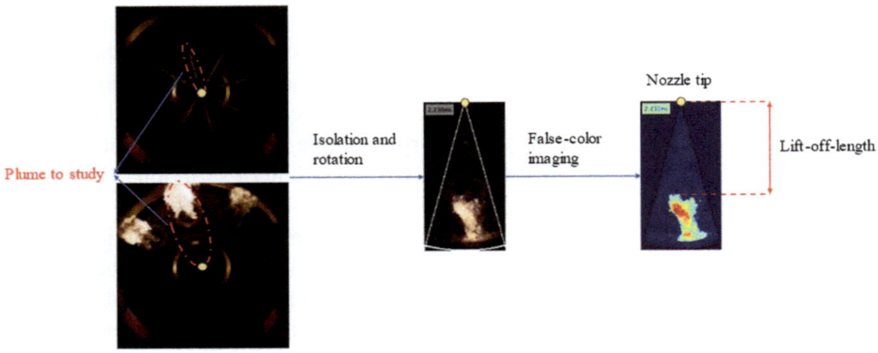

Fig. 3.3 Approaches for image processing

the shortfalls in design, fabrication, control, and others [15]. In this study, the start of injection (SOI) was determined as the first appearance of liquid fuel in each case.

The initial sequence of fuel spray includes liquid fuel delivery, liquid penetration, and vaporization. At 0.75 ms, the penetration length of the spray stabilized at approximately 22 mm from the injector tip. At 1 ms after injection, there was a clear reduction in penetration length owing to the cessation in liquid fuel delivery and the drop in plume momentum. At 1.08 ms, the image showed that all the residual liquid fuel disappeared owing to the liquid droplet atomization into a gaseous state which cannot be detected using scattered light.

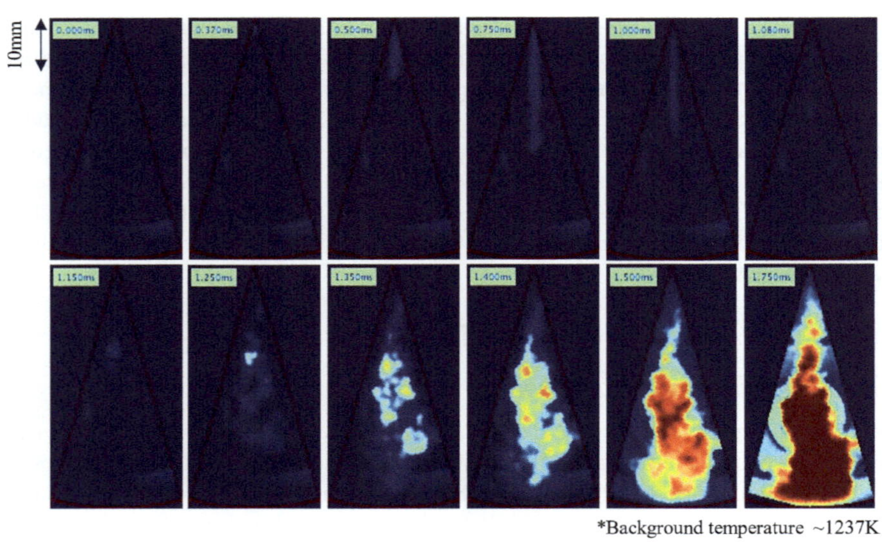

Fig. 3.4 The temporal sequence of the DME spray and ignition processes with 450 bar injection pressure and 0.6 ms injection duration

The first sight of a low-intensity flame was observed at 1.15 ms. The chemiluminescence became brighter as the flame propagated and grew, as shown at 1.25 and 1.35 ms. At 1.5 ms, significant chemiluminescence (high-intensity flame) that saturated the camera could be observed inside the low-intensity flame region, and the high-intensity flame grew rapidly and propagated downward and outward. By 1.75 ms, all the flame regions had transited into high-intensity flame.

A visual comparison of the reactive fuel sprays of DME and diesel at 1237 K is shown in Fig. 3.5a. The ignition and flame development processes for DME and diesel fuels were similar under these conditions. The first sight of autoignition pockets for both fuels is observed at 1.25 ms, shortly after liquid fuel delivery. Two stages of flame chemiluminescence were observed during the reaction processes. The first flames of low intensity were observed at a similar timing, yet DME showed a wider and more intense flame area. It is expected under these conditions that the enhanced volatility and rapid atomization of DME provided a shorter physical ignition delay [10]. The shorter ignition delay for DME allowed more time for flames to propagate and thus resulted in a wider flame area.

The flame intensities of both fuels increased as the flame propagated. The flame distribution for DME appeared more uniform than diesel. The intensity of the DME flame was evenly distributed while the diesel flame had a concentrated region of high-intensity flame. This difference can be attributed to the ignition of diesel fuels occurring at one location, and the flame propagated thereafter (refer to Fig. 5a at 1.25 ms). The ignition of DME spray appeared to occur independently at multiple locations. These flames propagated and eventually overlapped on a 2D viewing plane.

Figure 5b shows the comparison of ignition characteristics between DME and diesel fuels at 1168 K background temperature. Under lower background temperature, it was expected that a longer ignition delay and lower flame intensity would be observed owing to that the ignition delay increases with decreasing ambient temperature [16], and the proportion of pre-mixed burning flames increases as ignition delay prolongs [17]. At lower background temperatures, the low-intensity flame became more prominent, especially for DME fuel. Only a small region of high-intensity flame was observed at 2.25 ms for both fuels at this background temperature.

The flame for DME at the later stage of the ignition was much less intensified than diesel. This observation is owing to two factors. First, the prolonged ignition delay under low ambient temperatures allows more time for fuels to mix with the surrounding air. Second, the high volatility feature of DME results in a higher proportion of pre-mixed flame. Combining the abovementioned two factors, the combustion for DME fuel under these conditions exhibited a relatively more homogeneous combustion manner and high-temperature diffusion flame was avoided. However, diffusion flame was still predominant for diesel fuels under these conditions owing to the low volatility and resulting in high-temperature flame pockets.

At all tested conditions, DME exhibited a short ignition delay and lift-off length, as shown in Fig. 3.6. The variation of ignition between DME and diesel slightly reduced as the background temperature increased. A similar trend was apparent in the lift-off length.

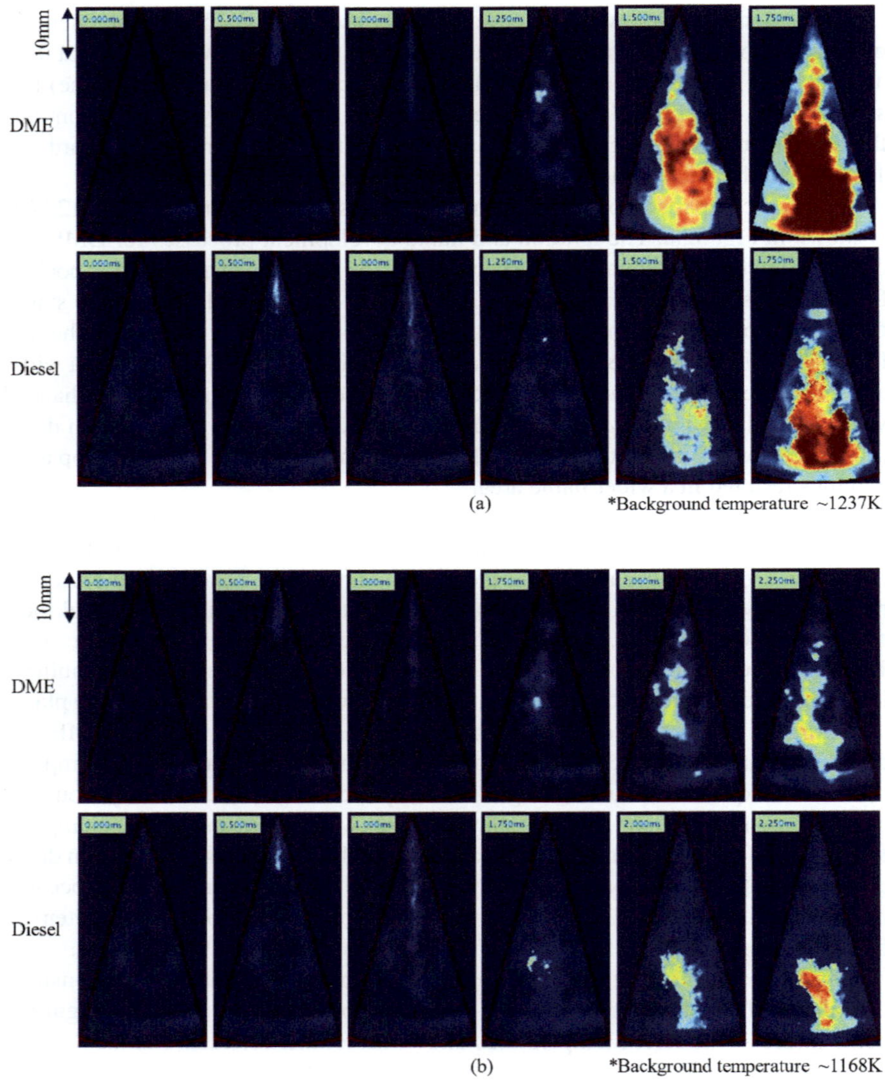

Fig. 3.5 A comparison of DME and diesel ignition characteristics at 1237 K, and 1168 K background temperature with 450 bar injection pressure and 0.6 ms injection duration

3.3.2 Combustion Characteristics

The combustion study conducted with a single-cylinder engine presents details of the fuel reactivity and combustion completeness. To facilitate a direct comparison between DME and diesel, both fuels were subject to matching engine operating

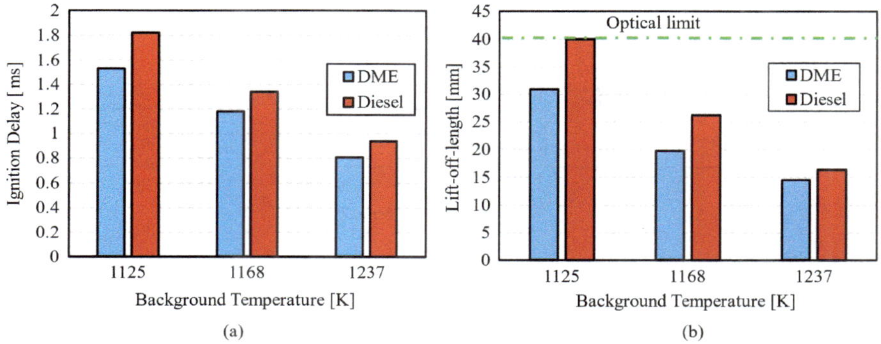

Fig. 3.6 The ignition delay and lift-off-length of DME and diesel at various background temperatures

and fuel injection pressure conditions. In relevance to the ignition characteristics (Sect. 3.1), the intake oxygen concentration was reduced from 21 to 15% through EGR. The in-cylinder pressure and apparent HRR are shown in Fig. 3.7. It is important to note that the injection timing and duration were adjusted to reach similar combustion phasing and engine load.

A shift in heat release distribution is observed with DME, wherein a second peak of heat release is apparent following the first. The comparatively lower energy density of DME requires an extended injection duration period under the same injection pressure. An overlap of fuel injection delivery and in-cylinder combustion is probable. As the early portion of the fuel is sufficiently vaporized and mixed with surrounding air

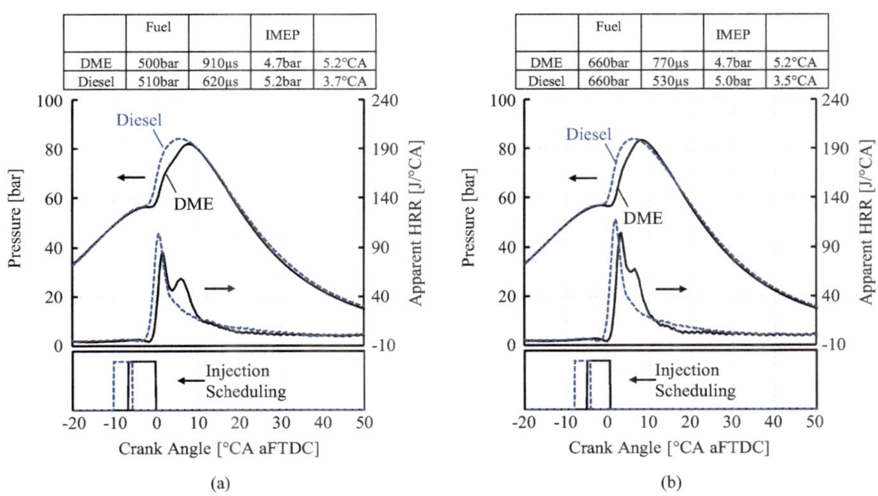

Fig. 3.7 In-cylinder pressure and apparent HRR with corresponding injection scheduling (below) at **a** 500~510 bar and **b** 660 bar injection pressure. The intake oxygen was 15~16%

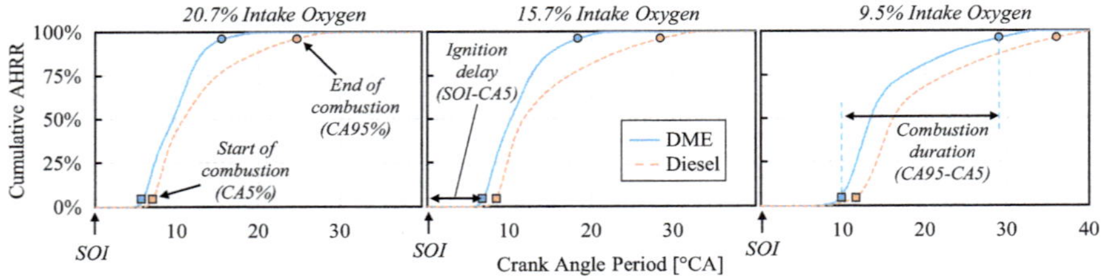

Fig. 3.8 The cumulative AHRR over the combustion period at 20.7, 15.7, and 9.5% intake oxygen. The traces are offset to begin at the commanded start of injection

to reach autoignition (first peak), the remaining regions of liquid fuel supply require a sufficient period before autoignition is reached, referred to as mixing-controlled combustion [18]. In stark contrast, diesel combustion shows a single peak of heat release under a shorter injection duration period.

An increase in injection pressure presents a second peak of heat release, albeit to a lesser intensity. The increase of injection pressure from 500 to 660 bar allowed the commanded injection duration pulse to decrease from 910 to 770 μs, albeit still considerably longer than diesel. A shorter overlap period of injection and combustion can realize a greater share of fuel–air regions sufficient for combustion at the onset of autoignition. In addition, the higher injection pressure enhances the atomization process and shortens the necessary mixing period to form regions with sufficient fuel–air for combustion [19]. The period from the start of injection to combustion, referred to as the ignition delay, was shortened by 1.1°CA.

The ignition delay period strongly depends on the charge reactivity, principally comprising the intake dilution rate and fuel property. A comparison of DME and diesel combustion processes at 660 bar at selected intake dilution levels is shown in Fig. 3.8. Each combustion history is referenced to its corresponding start of injection (SOI) command time. The onset of autoignition (noted with square markers) of DME occurs before diesel in all tested cases, ranging from 0.9~1.3°CA earlier. This result agrees with the reactive fuel spray ignition studies presented in Sect. 3.1.

The combustion duration is shorter with DME than with diesel. The initial heat release is sharper than diesel. This is most apparent at the lowest intake oxygen concentration, wherein diesel combustion shows a low slope from 0 ~ 7% cumulative AHRR. A similar normalized heat release trend is observed between DME and diesel, up to 50~70% cumulative AHRR. Beyond this combustion timing towards the end of combustion (noted with circle markers), an extended tail-end of combustion is apparent with diesel combustion. A similar trend can be observed with DME; however, it is most notable at the highest dilution rate of 9.5% intake oxygen. At 20.7% intake oxygen, the slight tail-end combustion occurs at ~85% heat released.

An enhanced charge reactivity of DME fuel presents advantages operating under highly diluted conditions, wherein oxygen depletion worsens the environment for ignition and combustion, as shown in Fig. 3.9. The comparison of engine load stability (Fig. 3.9a) shows the negligible difference among both fuels until heavy dilution wherein diesel combustion worsens while DME seems unaffected. The variation in pressure traces is showed through a 200 continuous cycle p–V trace, as shown in Fig. 3.9b. A thorough examination of the overlay traces reveals a slight variation in pressure after ignition, followed by higher variability near the peak in-cylinder pressure levels in diesel.

The principal obligation for intake dilution via EGR is to limit NOx emissions. The effectiveness of NOx emission suppression was apparent with DME and diesel combustion, as shown in the semi-log plot of Fig. 3.10a. Both fuels exhibit a monotonic exponential decay in NOx with lower intake oxygen concentration. However, the corresponding drop in combustion temperatures leads to combustion inefficiencies as unburned hydrocarbon species and carbon monoxide, as shown in Fig. 3.10b. Although NOx control is similar, the combustion inefficiencies are lesser with DME than with diesel. It is presumed that the primary cause of such combustion efficiency advantages is owed to the physical fuel properties of DME, namely the volatility, reactivity, and oxygen presence.

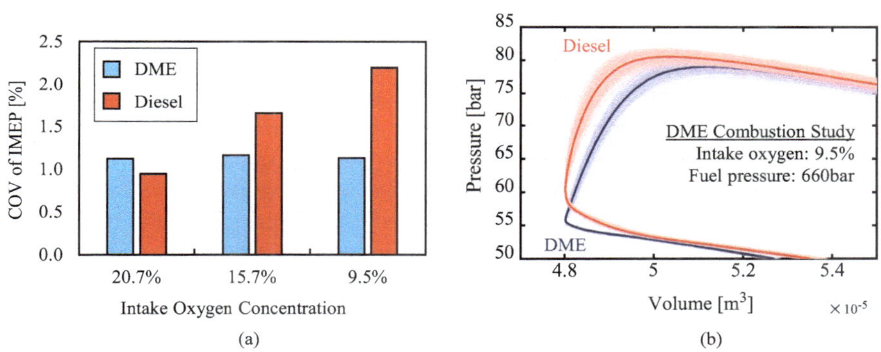

Fig. 3.9 **a** The variation of engine load at 20.7, 15.7, and 9.5% intake oxygen. **b** A p–V diagram of DME and diesel combustion at 9.5% intake oxygen with the average cycle trace (thick line) and 200 continuous cycles (thin line) superimposed

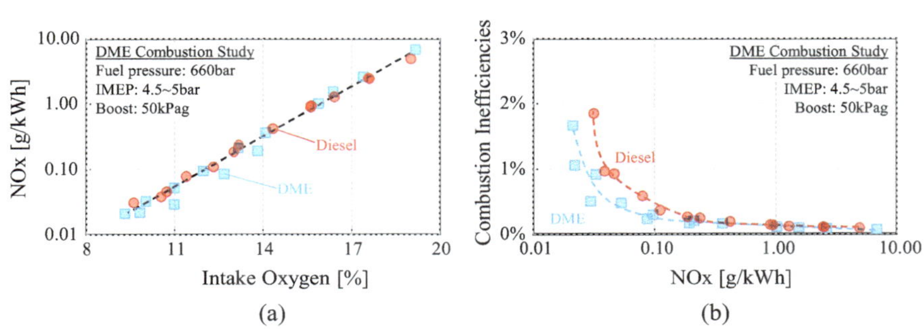

Fig. 3.10 **a** The influence of intake oxygen on NOx production. **b** The combustion inefficiencies under low NOx production

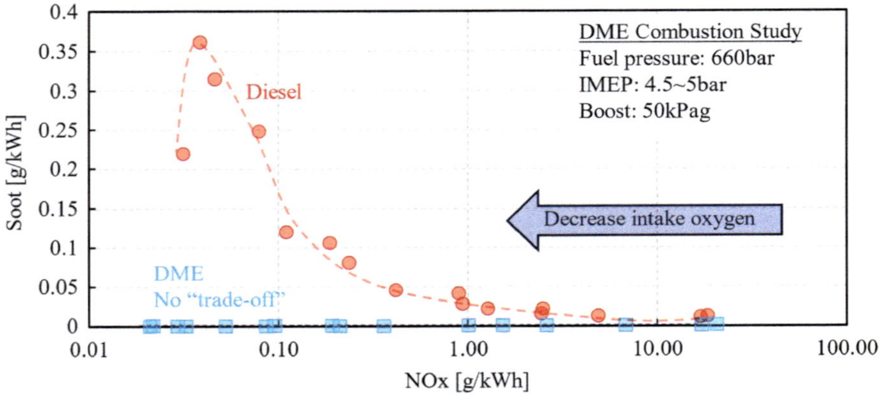

Fig. 3.11 The influence of intake oxygen on NOx-Soot production for DME and diesel

The reduction in NOx emissions could be effectively achieved via EGR. However, for diesel combustion the soot emissions deteriorated with lower intake oxygen concentration, as shown in the Fig. 3.11. As intake oxygen concentration decreases, the combustion temperature drops and leads to lower NOx emissions. However, the lowered oxygen concentration and lower combustion temperature make the soot oxidation more challenged and results in higher soot emissions. On the other hand, DME combustions showed ultra-low soot emissions ("mg/kWh" level) throughout the intake oxygen concentration. It is presumed that the primary cause of low soot emissions for DME is owed to the high borne-in oxygen content.

3.4 Conclusions

The ignition characteristics and combustion behaviour of DME were investigated using a constant-volume combustion chamber and a single-cylinder engine, respectively. Direct comparisons with diesel fuel were tested with matching conditions. For the ignition characteristics, two stages of ignition were observed for both DME and diesel fuels during the ignition processes. Low-intensity flame was first observed as a sign of the start of ignition and the flame transited into a high-intensity flame as it propagates. DME exhibited shorter ignition delay and lift-off-length compared to diesel, and the difference became more prominent as the background temperature decreased.

DME combustion showed an extended diffusion combustion period owing to its longer injection duration requirements. Improved combustion stability with DME combustion was clear, especially under heavy EGR diluted conditions. This became helpful on the pathway towards ultra-low engine-out NOx and PM emissions as DME proved a higher resistance to EGR and realized fewer combustion inefficiencies than diesel combustion.

In this study, direct comparison of DME and diesel ignition and combustion characteristics are conducted. For the ignition studies in the constant volume chamber, it will be helpful to understand the ignition characteristics if simultaneous schlieren imaging system could be achieved to observe the fuel vaporization effects. As for the combustion studies in the single cylinder research engine, the impact of injection pressure on combustion behaviors at different intake oxygen level for DME will be worthwhile to study.

Acknowledgements This research was partially supported by NSERC/IRC, NSERC/CRD, NSERC/RTI, NSERC/DG, CFI/ORF, Mitacs, and the University of Windsor. The authors thank the technologist at the University of Windsor, Ford Motor Company of Canada, and other OEM partners for their collaborative effort and technical support.

References

1. T. Johnson, Vehicular emissions in review. SAE Int. J. Engines **9**(2), 1258–1275 (2016). https://doi.org/10.4271/2016-01-0919
2. M. Zheng, G.T. Reader, J.G. Hawley, Diesel engine exhaust gas recirculation—a review on advanced and novel concepts. Energy Convers. Manag. **45**(6), 883–900 (2004). https://doi.org/10.1016/S0196-8904(03)00194-8
3. A.M. Kreso, J.H. Johnson, L.D. Gratz, S.T. Bagley, D.G. Leddy, A study of the effects of exhaust gas recirculation on heavy-duty diesel engine emissions. 981422 (1998). https://doi.org/10.4271/981422
4. M. Zheng, U. Asad, G.T. Reader, Y. Tan, M. Wang, Energy efficiency improvement strategies for a diesel engine in low-temperature combustion. Int. J. Energy Res. **33**(1), 8–28 (2009). https://doi.org/10.1002/er.1464
5. C. Arcoumanis, C. Bae, R. Crookes, E. Kinoshita, The potential of di-methyl ether (DME) as an alternative fuel for compression-ignition engines: a review. Fuel **87**(7), 1014–1030 (2008). https://doi.org/10.1016/j.fuel.2007.06.007
6. S.H. Park, C.S. Lee, Applicability of dimethyl ether (DME) in a compression ignition engine as an alternative fuel. Energy Convers. Manag. **86**, 848–863 (2014). https://doi.org/10.1016/j.enconman.2014.06.051
7. S. Sidhu, J. Graham, R. Striebich, Semi-volatile and particulate emissions from the combustion of alternative diesel fuels. Chemosphere **42**(5–7), 681–690 (2001). https://doi.org/10.1016/S0045-6535(00)00242-3
8. Y. Sato, A. Noda, T. Sakamoto, Y. Goto, *Performance and Emission Characteristics of a DI Diesel Engine Operated on Dimethyl Ether Applying EGR with Supercharging* (2000-01-1809) (2000). https://doi.org/10.4271/2000-01-1809
9. Y. Sato, S. Nozaki, T. Noda, *The Performance of a Diesel Engine for Light Duty Truck Using a Jerk Type In-Line DME Injection System* (2004-01-1862) (2004). https://doi.org/10.4271/2004-01-1862
10. H. Teng, J.C. McCandless, J.B. Schneyer, Compression ignition delay (physical + chemical) of dimethyl ether—an alternative fuel for compression-ignition engines. SAE Int. J. Fuels Lubr. **112**, 377–389 (2003)
11. D.W. Gill, H. Ofner, C. Stoewe, K. Wieser, E. Winklhofer, M. Kato, T. Yokota, J. Weber, *An Investigation into the Effect of Fuel Injection System Improvements on the Injection and*

Combustion of DiMethyl Ether in a Diesel Cycle Engine (2014-01-2658) (2014). https://doi.org/10.4271/2014-01-2658
12. X. Yu, S. LeBlanc, N. Sandhu, J. Tjong, M. Zheng, Combustion control of DME HCCI using charge dilution and spark assistance. Proc. Inst. Mech. Eng. Part J. Automob. Eng. 095440702211033 (2022). https://doi.org/10.1177/09544070221103361
13. Y. Mitsugi, D. Wakabayashi, K. Tanaka, M. Konno, High-speed observation and modeling of dimethyl ether spray combustion at engine-like conditions. SAE Int. J. Engines **9**(1), 210–221 (2015). https://doi.org/10.4271/2015-01-1927
14. S. LeBlanc, X. Yu, L. Wang, M. Zheng, Dimethyl ether to power next-generation road transportation. Int. J. Automot. Manuf. Mater. (2023)
15. J.B. Heywood, *Internal Combustion Engine Fundamentals* (McGraw-Hill, New York, 1988)
16. L.-M. Malbec, W.E. Eagle, M.P.B. Musculus, P. Schihl, Influence of injection duration and ambient temperature on the ignition delay in a 2.34L optical diesel engine. SAE Int. J. Engines **9**(1), 47–70 (2015). https://doi.org/10.4271/2015-01-1830
17. S. LeBlanc, L. Jin, A. Bastable, X. Yu, J. Tjong, M. Zheng, An optical investigation into the reactive fuel spray of high-pressure DME and ethanol. Proc. Int. Symp. Diagn. Model. Combust. Intern. Combust. Engines **10**, C1-4 (2022). https://doi.org/10.1299/jmsesdm.2022.10.C1-4
18. C.E. Dumitrescu, C. Polonowski, B.T. Fisher, A.S. Cheng, G.K. Lilik, C.J. Mueller (ed.), An experimental study of diesel-fuel property effects on mixing-controlled combustion in a heavy-duty optical CI engine. SAE Int. J. Fuels Lubr. **7**(1), 65–81 (2014). https://doi.org/10.4271/2014-01-1260
19. M.K. Le, S. Kook, Injection pressure effects on the flame development in a light-duty optical diesel engine. SAE Int. J. Engines **8**(2), 609–624 (2015). https://doi.org/10.4271/2015-01-0791

Chapter 4
A Critical Appraisal of Batteries with Metal Phosphate Among Commercial Batteries

Figen Balo and Lutfu S. Sua

Abstract Power storage devices have been proven to be beneficial for electricity generation. A battery device is structured with an electronic element outside the cell and an electrolyte that transfers ionic element of the chemical reaction within the battery. On discharge, the output is an exterior electrical current at a certain voltage. When implementing a charging voltage and current, chemical reaction of a rechargeable battery must be reversible. Safety, energy density at a specific energy output, storage efficiency, shelf and cycle life, and fabrication cost are among the critical factors of rechargeable batteries. During charge and discharge, Li-ion power units are rechargeable battery types in which Li-ions flow from the negative to the positive electrode. Cost, performance, safety, and chemistry parameters change across Li-ion battery types. In consumer electronics, Li-ion batteries are widely used. Li-ion batteries, unlike Li main cells which are disposable, utilize an intervened Li combination as the electrode material rather than metallic Li. They have one of the best weight-to-power rates, no memory impact, a low self-discharge ratio, a considerable open circuit voltage, and slow loss of charge while not in utilization, making them one of the most comprehensive types of rechargeable power units for portable devices. Due to their high power density, Li-ion batteries are becoming more popular in military, aerospace implementations, and electric cars. In this study, an analysis is made using a multi-attribute decision-support technique for the rechargeable batteries available in the commercial market. A critical evaluation is carried out to determine the most efficient rechargeable battery by using the important parameters in terms of their efficiencies. Research findings on general and current developments in lithium-ion (Li-ion) batteries with metal phosphate were compiled and its place among the suitable rechargeable batteries is determined with the help of mathematical analysis. The research aims to guide the relevant people on the selection of rechargeable

F. Balo (✉)
Department of METE, Engineering Faculty, Firat University, Elazığ, Turkey
e-mail: figenbalo@gmail.com

L. S. Sua
Department of Management and Marketing, Southern University and A&M College, Baton Rouge, USA

D. S-K. Ting and A. Vasel-Be-Hagh (eds.), *Engineering to Adapt*, Springer Proceedings in Energy, https://doi.org/10.1007/978-3-031-47237-4_4

batteries. The technical properties of batteries with metal phosphate among commercial batteries were investigated. A criteria set is developed for the assessment of rechargeable batteries. Six rechargeable battery alternatives are investigated.

Keywords Li-ion battery · Rechargeable battery · Efficiency · Battery with metal phosphate

4.1 Introduction

Li-ion (lithium-ion) batteries involve high technology that operate Li-ions as it is a significant element of electrochemistry. In the anode, Li-atoms are separated and ionized from their electrons in the course of a discharge cycle. From the electrolyte to the cathode, the Li-ions work from the anode. Here they remix by electrically neutralizing through electrons. These are not sufficient to change by a micro permeable divider that divides the cathode and anode. Due to the minimal size of lithium, Lithium-ion batteries can have extreme charge storage and voltage per unit volume and mass [1, 2]. Electrodes in Li-ion batteries can be obtained from a variety of materials. The most popular compound is the cathode (Li-CoO_2) and anode (graphite), which are obtained in portable electronic tools such as laptop computers and cell phones. Li-MgO_2 (utilized in electric and hybrid cars) and Li-Fe-PO_4^{3-} are two other materials of the cathode. Since it is an electrolyte, ether (a group of organic combinations) is generally utilized in Li-ion batteries [1]. A Li-ion battery, also known as lithium-ions, is a type of rechargeable power unit constructed of cells in which Li-ion moves from the negative to the positive electrode by an electrolyte in the course of discharge and back again throughout charging. Lithium-ion cells utilize an intervened Li combination as the positive electrode material and characteristically graphite as the negatory electrode material. Lithium-ion power units have a good power intensity, no memory impact, and a lower self-discharge ratio. Cells can be planned according to energy density or power [3]. However, since they contain inflammable electrolytes, they can create explosions and fires if incorrectly charged or damaged.

Lithium-ion power units have a few advantages over other high-quality rechargeable power units (Ni-metal-hydride or Ni–Cd). They have the maximum power density of any power unit available nowadays (670–250 Wh/L or 265–100 Wh/kg). In addition, Lithium-ion power unit cells can deliver up to 3.60 V, which is 3 times higher than Ni-MH or Ni–Cd technologies. This implies that they can provide large amounts of current for high-energy implementations. Lithium-ion batteries also require less maintenance, as they do not need planned cycling to sustain power unit life [4].

Lithium-ion batteries have a memoryless impact, which is a negative operation in which a power unit can "remember" a lower capability after partial charge/discharge repeated cycles. This is a benefit over Ni-MH and Ni–Cd, both of which show the same impact. Lithium-ion batteries also have a less monthly self-discharge ratio of

0.02–0.015. Since they do not include toxic Cd, they are easier to keep under control than Ni–Cd batteries [5]. Due to these benefits, Lithium-ion batteries have exceeded Ni–Cd batteries to become the market pioneer in transportable electronic tools (like laptops and smartphones). Lithium-ion batteries are also utilized to actuate electric mechanisms in several aero-space implementations, particularly in the newer and more environment-friendly planes, where weight is an important cost element.

Much of the Lithium-ion industry's promise concerning clean power stems from its potential implementations in power unit automobiles. Nowadays, the most popular electrical vehicles are powered by Li-ion batteries [1, 5]. Despite their industrial potential, Lithium-ion batteries have plenty of defects, especially from a security viewpoint. Lithium-ion batteries superheat easily and can be damaged at high voltages. This can conclude in combustion and thermal runaway in some cases. This has caused significant topics, containing the plane fleet succeeding records' grounding of aboard power unit blazes. Due to the risks associated with these batteries, several transportation companies reject shipping batteries collectively by aircraft. Lithium-ion power units necessitate safety systems to restrict internal pressures and voltage, which can raise weights and, sometimes, limit efficiency.

Lithium-ion power units, such as all power units, age, which implies they frequently fail several years later and lose potential. Another barrier to extensive acceptance is their high expense, which is roughly 40% higher than Ni–Cd. Giving address to these topics is a critical element of current technological search. Eventually, despite having a greater power intensity than batteries of other kinds, Li-ion batteries are still about one hundred times less power intense than benzine (which includes 8760 Wh/L through volume or 12.700 Wh/kg through mass) [6].

Batteries are the most conventional type of storage device with ecological advantages, fuel flexibility, and other benefits [7]. However, numerous producers are offering a variety of technologies and features even for rechargeable batteries. The following characteristics have been identified by Akindele and Rayudu: discharge time, sector energy rating, discharge losses (daily), cycling capacity, proper storage duration, power and energy density, lifetime (yearly), and round-trip performance (%) [8]. As a result, a new decision-support question has appeared, centered on selecting the best energy storage alternatives from various options. This necessitates the use of multiple-criteria decision-support systems. While large-scale industries can use a consulting firm's assistance in decision support, this is a difficult task for small-scale businesses or households.

The multi-attribute decision analysis methodologies are frequently utilized to address decision-making issues. Concerning material choice, diverse multi-attribute decision analysis methodologies were utilized such as Fuzzy-AHP and AHP, Fuzzy-TOPSIS and TOPSIS, SMARTER, VIKOR, COMET, and DEMATEL methodologies. An itemized review of such methodologies and many other multi-attribute decision analysis methodologies can be obtained [9, 10].

Operations research has a subdiscipline called multi-attribute decision analysis. Its goal is to simplify complicated decision-support issues into comprehensive, simple, and actionable types for the experts [11]. The assessment task depends on criterion comparisons that involve attribute trade-offs [12]. It can treat incomparable objectives like profit, cost, safety, and efficiency, and as a result, provide wide applicability, specifically in complicated sustainability problems [13]. The hierarchy of multi-attribute decision analysis is made up of four parts: attributes and criteria, alternatives, weights, and efficiency information.

In the context of this paper, multi-attribute decision analysis refers to a method of evaluating alternatives by combining battery data, decision attributes, and expert precedences. When decision-makers face this type of problems, the necessary element that distinguishes the alternatives is the decision preferences of experts [12]. Li et al. developed a decision-making system to resolve the energy storage sector's multiple objective optimization decision-support questions, taking into account three objectives: maximizing technic suitability, minimizing expense, and minimizing ecological effect [14].

Olabi suggested multi-criteria choice attributes for energy storage mechanisms, containing energy source usability, energy application, and requirement, energy storage performance, infrastructure, and cost [15]. The conventional technical-financial analysis methodology to comprehensively assess the energy safekeeping area mechanisms of Li-ion and Pb-acid power units is used by Dhundhara et al.. All the same, the choice of battery energy safekeeping area system is a multiple-criteria issue. The conventional technical-financial approach cannot completely mirror the real condition of battery energy safekeeping area units due to its sole-sidedness. Thus, the multiple-criteria assessment methodology should be accepted to select and assess battery energy safekeeping area systems [16]. Zhang et al. presented 7 kinds of power unit mechanisms, related to the material, electro-chemistry efficiency, and expense of power units. It was indicated that the capacity and performance of great-scale energy safekeeping area mechanisms are centered on power condition and battery [17]. The studying foundations, voltage, operating temperature, storage capacity, cycling times, and implementation of energy storage power units are presented by Poullikkas and Nikolaidis.

An extensive analysis was performed regarding the electrochemical and mechanical industries and capital cost information that was utilized in each of the power safekeeping area mechanisms to assess the feasibility of these mechanisms[18]. The batteries' diverse kinds for great-scale electricity storage, especially the current great-scale battery energy safekeeping area system in the process and other kinds of great-scale power safekeeping area mechanisms outlined by Poullikkas. The approach represents that Na–sulfur power units are appropriate for great-scale power safe-keeping area mechanisms, while the flux power units, especially the vanadium redox flux power units are appropriate for little-scale power safekeeping area mechanisms [19]. Yuqing et al. supplied the attributes and methodologies for assessing power unit extent in power safekeeping area mechanism and their implementations in different renewable power mechanisms, to illuminate the power unit's primary indicator and to achieve the equilibrium between the additional cost and the technic improvements [20].

In this research, the AHP methodology is performed to select the most beneficial power unit sector among six diverse sectors concerning ten attributes. The binary sorting of each power unit technology according to each of the criteria is confirmed by assessing the diverse values for each of the alternative materials according to each of the criteria. The AHP methodology provides the best benchmark among the diverse power unit sectors.

4.2 Analyzed Rechargeable Batteries and Parameters Used in the Analysis

The specific energy intensity is the amount of power available to the working electrode material per unit of weight. In a whole discharge loop, it is the product of the operating voltage and the specific capacity. Because both voltage and current can fluctuate during a discharge loop, the specific energy obtained is computed by adapting the voltage over time. For a rechargeable power unit, the discharge time is proportional to the maximal and minimal voltage thresholds and is determined by the active materials' availability and evasion of an irreversible form [21]. For electric cars, great specific energy intensity power units are gaining popularity as potential energy resources. Owing to their far greater theoretic specific energy intensity than traditional power units, Li-air batteries are the most promising mechanism [22]. A cell's internal impedance determines its current carrying capacity. High currents are possible due to low internal resistance [21].

The physical properties of the electrolyte influence internal resistance; the lower the impedance, the lower the granulous extent of the electrolyte material. In a milling process, granule extent is regulated through the cell producer [23]. A rechargeable power unit's cycle life is characterized as the number of recharge/charge loops a second power unit can conduct before its potential drops to 0.80 of what it was at first. This ranges from 1200 to 500 loops [21]. Supercharge tolerance is high, and performance is safer. Continual charging above 4.30 V would either cause an explosion or fire or degrade power unit efficiency (like life-cycle) [21]. Self-discharge is a battery event where inside chemical treatments decrease the battery store charge with no relationship among any outside circuitry or electrodes. The self-discharge shortens the batteries' life and creates them to charge lower than a whole charge when used [24]. In electronics, the voltage of cut-off is the voltage at which a power unit is completely discharged and more discharge could create damage. When the cut-off voltage is reached, several electronic tools, like cell telephones, will close automatically. The discharge cutoff voltage is a significant element in deciding the lithium battery's lifetime. As a result, the meaning of this concept should be clarified.

The term"Li-battery" specifically refers to a lithium-ion power unit. If a lithium battery has reached its cut-off voltage, it is no longer proper for discharge. When the voltage accesses a specific level, the power unit is devastated, and the Li power unit's partial capability is irreversible. A sole Li power unit's voltage rated is 3.70 V, and the cut-off voltage is 2.75 V. However, in the latter scenarios, companies may increase the cut-off voltage to 3.00 V; one is related to special electric tools and functioning, and the other is to make battery use safer. By contrast, manufacturers may decrease the cut-off voltage to 2.40 V or 2.50 V. There is no voltage cut-off lesser than 2.40 V [25]. The cut-off discharge voltage denotes the point at which the discharge is complete. In practice, this voltage represents the maximum capacity of the power unit. One of the most efficient policies for increasing the Li-ion batteries' energy density is to increase the charge cut-off voltage. Some portable equipment does not fully utilize a power unit's low final voltage spectrum. The energy to the device is cut off before a significant portion of the battery life is consumed. An extreme cut-off voltage is common. For instance, a specific mobile phone brand with a sole-cell Li-ion power unit shuts down at 3.30 V. The Li-ion battery can be discharged to 3.00 V (at room temperature) or lesser.

All the same, a discharge to 3.30 V uses approximately 0.92–0.98 of the capacity [6]. Significantly, in the event of Li-ion power units, which are utilized in the portable electronics' huge plurality nowadays, a voltage cut-off beneath 3.20 V can cause a chemical imbalance in the cell, resulting in a decreased power unit lifetime. As a result, electronics producers utilize greater cut-off voltages, eliminating the requirement for users to purchase power unit changings before other malfunction systems in a tool kick in [26].

In the course of charging, an outside electrical energy resource (the charging circuitry) implies an overvoltage (a higher voltage than the identical polarity's power unit generates), causing a charging current to flux within each of the cells from the plus to the negatory electrode, e.g., in the opposite way of a discharge current underneath regular circumstances. In an operation called intervening, Li-ions immigrate from the plus to the negatory electrode and occur embedded in the electrode material with porous [27]. When a battery is stored for an extended period, the little current draw of the conservation circuit system may discharge the power unit under its close voltage. The regular chargers may then be rendered ineffective because the power unit administration mechanism may keep a report of this charger (or power unit) "malfunction". Most Li-ion cells cannot be safely charged below 0 °C [28], as this can result in lithium covering the cell's anode, which can create complexities like inside short-circuitry ways. Discharging the battery under its theoretical voltage can cause operations that destroy or damage the battery. As a result, batteries should include electronics that alert the consumer and turn off the apparatus before such a whole discharge occurs [29]. The peak load current is the maximal current at which the power unit can be continuously drained.

The power unit producer generally specifies this restriction to avoid extreme drain ratios from damaging the power unit or reducing its capability. The voltage of the battery varies depending on the discharge current used (load). If the discharge operating temperature range is −2060 °C, Li batteries with 2.75 V cut-off voltage can continue to discharge. All the same, a drain with a voltage of less than 2.5 V is not permitted because drain loss from 2.50 to 2.75 V may be superficial, but a drain less than 2.50 V will result in critical battery harm [25].

Some rechargeable batteries used in portable electronic devices may be dangerous metal polluters in the ecology. These metal polluters can have a negatory effect on human health and ecological quality, especially in areas where solid waste recycling, sorting, and collection infrastructure is lacking [30]. Coulombic performance, also known as current performance or faradaic performance, refers to the charge performance with which electrons in batteries are converted. Coulombic performance is defined as the rate of the whole charge sorted from the power unit to the whole charge put inside the power unit over the course of a whole loop. Lithium-ion rechargeable batteries have one of the maximum Coulombic performance evaluations. It has a better performance than 0.99. This is solely feasible if the power unit is charged at a cool temperature and moderate current. Due to the losses because of heat and charge acceptance, ultra-fast charging lowers the Coulombic performance, as does ultra-slow charging with self-discharge [31].

Most of the diurnal actions would be unfeasible to fill without the capability to recharge rechargeable power units. NiCd, Pb-acid, Li-ion, and Ni-MH are the most popular rechargeable power units. A summary of their features is given below [32–36].

- Pb-acid is the most traditional rechargeable power unit mechanism. Pb-acid power units are reasonably priced and tough. Additionally, they have a restricted loop count and lower specific energy. Pb-acid power units are utilized in golf carts, wheelchairs, personnel carriers, backup energy providers, and emergency lighting. Pb is toxicological. In a dump, it should never be kept under control.
- Ni–Cd is a well-understood and mature metal that is utilized in implementations necessitating long service time of life, extreme temperatures, and good discharge current. Ni–Cd batteries are amongst the most long-lasting and durable; they are the sole chemistry that permits ultrafast charging with minimum strain. Medical devices, energy tools, UPS, and aeronautics are the most general implementations. Ni–Cd is being compensated with other chemical materials owing to ecological considerations, but it holds its trend in air-craft owing to its high-security report [37].
- Ni-MH is utilized in place of Ni–Cd because it includes solely mildly toxicological metals and has a greater specific energy. Ni-MH is utilized in hybrid vehicles, medical devices, and sectoral implementations. For user utilization, Ni-MH is advanced in AAA and AA cells too.
- Lithium-ion power units compensate Pb and Ni-sourced power units in most implementations. Lithium-ion needs conservation circuitry because of security apprehensions. It is more costly than the majority of other power units, but the high loop count and less upkeep lower the expense of every loop when crosschecked with most of the other chemical matters [38].

Table 4.1 crosschecks the features of four generally utilized rechargeable power unit mechanisms, displaying mean efficiency evaluations at the publication time. Lithium-ion power units are classified according to their working materials, which are manganese, phosphate, titanate, and cobalt. The known Li-ion and plastic, which receive their titles from the separator mechanism and one electrolyte, are not on the list. Most are hybrids that share performance with other Li-ion batteries [39].

Another type of unit not on the list is the rechargeable Li-metal power unit, which has the potency to become a power unit selection with radically good specific power and high specific energy once the safety issues are resolved. This Table solely

Table 4.1 Characteristical properties of commonly utilized rechargeable batteries

Specifications	Lead acid	Ni–Cd	Ni-MH	Li-ion		
				Cobalt	Manganese	Phosphate
Charge time (about)	12 h	1.5 h	3 h	3 h	1.5 h	1.5 h
Internal resistance	Very low	Very low	Low	Moderate	Low	Very low
Specific energy (Wh/kg)	40	63	90	200	125	55
Cycle life (80% DoD)	250	1000	400	750	750	1500
Self-discharge/ month (room/ temp)	5%	20%	30%	Small 5%		
Overcharge tolerance	High	Moderate	Low	Low. No trickle charge		
Peak load current (best result)	5 °C 0.2 °C	20 °C 1 °C	5 °C 0.5 °C	2 °C small 1 °C	Big 30 °C small 10 °C	Big 30 °C small 10 °C
Cell voltage (nominal)	2 V	1.2 V	1.2 V	3.6 V	3.7 V	3.25 V
Discharge cutoff voltage (V/cell 1 °C)	1.75 V	1.00 V		2.75 V		2. 50 V
Charge cutoff voltage (V/cell)	2.40 Float 2.25	Full charge detection by voltage signature		4.20 typically some go higher		3.60 V
In use since	The late 1800s	1950	1990	1991	1996	1999
Safety requirement	Thermally stable	Thermally stable, fuse protection		Protection circuit mandatory		
Maintenance requirement	4.5 months (loping chg.)	Full discharge every 90 days when in full use		Maintenance-free		
Toxicity	Very high	Very high	Low	Low		
Discharge temperature	−20 to50 °C	−20 to 65 °C		−20 to 60 °C		
Charge temperature	−20 to50 °C	0 to 45 °C		0 to 45 °C		
Cost	Low	Moderate		High		
Coulombic efficiency	−90%	−70% slow charge −90% fast charge		99%		

discusses transportable power units and does not include big mechanisms resembling refineries. This study aims to reveal the order of effectiveness of Li batteries among the types of Li batteries and other portable batteries. Figures depend on average commercial battery evaluations at the publication time. Especially power units with higher-than-average evaluations are not included.

Characteristical properties of commonly utilized rechargeable batteries are given in Table 4.1 [40].

4.3 Methodology

Analytic Hierarchy Process methodology is utilized to identify the diverse criteria weights that are taken into consideration. The most utilized attributes for rechargeable battery choice are charge time, cycle life, specific energy, cell voltage, self-discharge, overcharge tolerance, toxicity, peak load current, discharge cutoff voltage, and cost.

Introduced in 1980, AHP is a decision-making method based on multiple criteria. The following steps must be taken when applying AHP [41]:

Stage 1: A decision-support matrix is created and weights (priorities) are assigned to the attributes. Diverse application attributes are cross-checked based on the desired objective to determine their weights. A panel of experts is formed from the industry and academics to evaluate the alternatives based on the list of criteria. An equal number of experts from industry and academics (10 each) have participated in a short survey voluntarily to provide their pairwise comparison of the selected criteria. The sample statistics of the experts who participated in the study show that the year of experience in the field is 7.1 on average, varying between 5 and 15 years.

Each entry in this matrix is on a 1–9 scale, which is shown in Table 4.1. Furthermore, the benchmark matrix is dependent on the experts' assessment of the notional significance of various attributes. The attributes in row i [i = 1, 2, …. n] is sorted in relation to each of the n attributes.

G represents the pairwise comparison assessment matrix (n x n). g_{ij} [i,j = 1, 2, …. n] displays the notional significance of the attribute i in relation to the attribute j for an imparted factor from the matrix G. When a criterion is compared to itself, the entry input in the matrix is assigned the value "one". As a result, the primary transversal inputs of the binary benchmark matrix are overall "one". The binary benchmark's prioritization scale is given in Table 4.2 [41].

Stage 2: the matrix is normalized through the division of each entry from the column of the matrix by its column total.

Stage 3: the mean of each normalized matrix row is calculated. These row averages are used to calculate the nominal weights w_i [i = 1,2, …n] prioritizes concerning each criteron.

Stage 4: the level of accuracy is confirmed. If "w" is the relative weight column vector, the benchmark matrix G is coherent if: $G \cdot W = n \cdot W$

Table 4.2 The binary benchmark's prioritization scale

Nominal significance intensity	Definition
2–4-6–8	The intermediate-level decision among 2 adjacent decisions
9	Extremely significant
7	Too strongly significant
5	Strongly significant
3	Moderate significant
1	Equal significant

Matrix X is a column vector with n dimensions. It is the weighted average of each of the alternatives when each criterion is taken into account;

$$G := \begin{bmatrix} 1 & g_{12} & \cdots & g_{1n} \\ g_{21} & 1 & \cdots & g_{2n} \\ \cdots & \cdots & \cdots & \cdots \\ g_{n1} & g_{n2} & \cdots & 1 \end{bmatrix} \tag{4.1}$$

$$\mathrm{X} := \mathrm{GW} = \begin{bmatrix} 1 & \mathrm{g}_{12} & \cdots & \mathrm{g}_{1\mathrm{n}} \\ \mathrm{g}_{21} & 1 & \cdots & \mathrm{g}_{2\mathrm{n}} \\ \cdots & \cdots & \cdots & \cdots \\ \mathrm{g}_{\mathrm{n}1} & \mathrm{g}_{\mathrm{n}2} & \cdots & 1 \end{bmatrix} \begin{bmatrix} \mathrm{w}_1 \\ \mathrm{w}_2 \\ \cdots \\ \mathrm{w}_\mathrm{n} \end{bmatrix} = \begin{bmatrix} \mathrm{c}_1 \\ \mathrm{c}_2 \\ \cdots \\ \mathrm{c}_\mathrm{n} \end{bmatrix} \tag{4.2}$$

where; G: decision matrix and X: factor weights

Stage 5: Evaluate the accuracy valuations for the group of alternatives.

Stage 6: Eigen valuation computation:

$$\lambda_{\max} = \frac{\sum_{i=1}^{i=n} CV_i}{n} \tag{4.3}$$

Here, "n" is the number of factors to be compared and λ are eigenvalues.

Stage 7: The accuracy index (CI) is determined as follows. It should be considered that the output quality of the AHP methodologies is directly proportional to the accuracy of the binary benchmark judgments;

$$CI = \frac{(\lambda_{\max} - n)}{n - 1} \tag{4.4}$$

Stage 8: Calculate the arbitrary inaccuracy index [41] (Table 4.3).

Stage 9: Compute the accuracy rate (CR), which is specified as:

Table 4.3 Arbitrary consistency index (RI) for various matrix extents

Matrix size (n)	1	2	3	4	5	6	7	8	9	10
RI	0.00	0.00	0.58	0.90	1.12	1.24	1.32	1.41	1.45	1.49

$$CR = \frac{CI}{RI} \tag{4.5}$$

For the degree of comparison accuracy to be favorable, the accuracy rate should be 0.10. If the accuracy ratio is bigger than 0.10, the assessment process must be ingeminated.

4.4 Results and Discussion

The AHP technique is used to find the optimum alternatives in various fields. This study illustrates the applicability of the AHP in choosing rechargeable batteries. In addition, a few thorough case studies of selecting and assessing these battery alternatives and elements by using the AHP approach are described as a reference for the reader in evaluating, selecting, and manufacturing correct rechargeable batteries. The selection process analyzed significant parameters based on the batteries and certain characterization properties of the area are researched at the outset of the research.

Ten different factors are considered for evaluating the rechargeable battery alternatives and these factors are crosschecked based on the supplied comparison measurement. The emerging decision-support matrix is presented in Table 4.4.

Utilizing binary comparisons of the listed factors, the following relative weights are computed based on professional assessments. The weights are displayed in Fig. 4.1. As the figure displays, the peak load current, toxicity, and specific energy are evaluated to be the most significant factors in evaluating the alternatives.

The methodology's final stage involves evaluating the alternatives based on each factor and multiplying the values with the factor weights. The resulting values are called priority values. In Table 4.5, these values are presented. The last row displays all priority values of each alternative.

Depending on the values, Fig. 4.2 demonstrates the attractiveness of the selected batteries. Figure 4.2 indicates that *Co* is the option with the maximum point, with a value of 0.0563.

Table 4.4 Decision matrix

		Specific energy	Cycle life	Charge time	Overcharge tolerance	Self-discharge	Cell voltage	Discharge cutoff voltage	Peak load current	Toxicity	Cost	Weights
Specific energy	1	1	1	5	7	3	2	2	0.5	1	6	14,1
Cycle life	2	1	1	3	0.5	6	0.5	0.5	0.33	0.17	2	8.54
Charge time	3	1/5	1/3	1	0.33	7	0.17	0.17	0.14	0.17	2	6.57
Overcharge tolerance	4	1/7	2	3	1	2	0.33	0.33	0.17	0.17	2	5.60
Self-discharge	5	1/3	1/6	1/7	1/2	1	2	3	3	1	4	9.80
Cell voltage	6	1/2	2	6	3	1/2	1	2	0.33	2	0.17	9.64
Discharge cutoff voltage	7	1/2	2	6	3	1/3	1/2	1	0.5	1	4	9.25
Peak load current	8	2	3	7	6	1/3	3	2	1	2	3	15.9
Toxicity	9	1	6	6	6	1	1/2	1	1/2	1	8	14.8
Cost	10	1/6	1/2	1/2	1/2	1/4	6	1/4	1/3	1/8	1	5.68

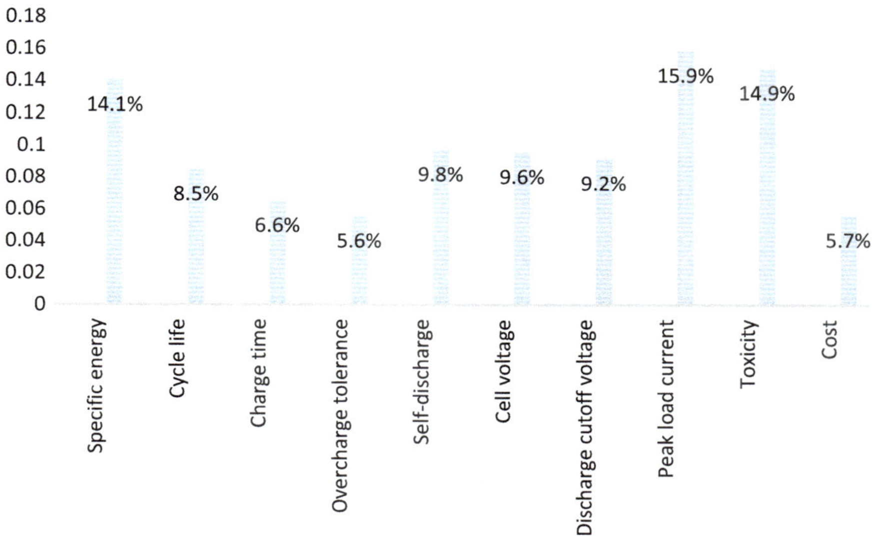

Fig. 4.1 Factor weights

Table 4.5 Priority values

	Pb Acid	NiCd	NiMH	Cobalt	Manganese	Phosphate
Specific energy	0.009	0.014	0.020	0.045	0.028	0.024
Cycle life	0.005	0.018	0.007	0.014	0.014	0.028
Charge time	−0.035	−0.004	−0.009	−0.009	−0.004	−0.004
Overcharge tolerance	−0.019	−0.012	−0.006	−0.006	−0.006	−0.006
Self-discharge	−0.008	−0.031	−0.046	−0.005	−0.005	−0.005
Cell voltage	0.013	0.008	0.008	0.023	0.024	0.021
Discharge cutoff voltage	0.014	0.008	0.008	0.022	0.022	0.020
Peak load current	−0.009	−0.035	−0.009	−0.003	−0.052	−0.052
Toxicity	−0.050	−0.050	−0.012	−0.012	−0.012	−0.012
Cost	−0.004	−0.008	−0.008	−0.012	−0.012	−0.012
	−0.0833	−0.0917	−0.0468	0.0563	−0.0042	0.0002

4.5 Conclusions

Technical assessment plays a critical role in the assessment of technological and scientific accomplishments. The industry is interested in alternative ways of technological and scientific accomplishments which have a significant effect on efficiency, the economy, and several other ways. The technical specification of power units with

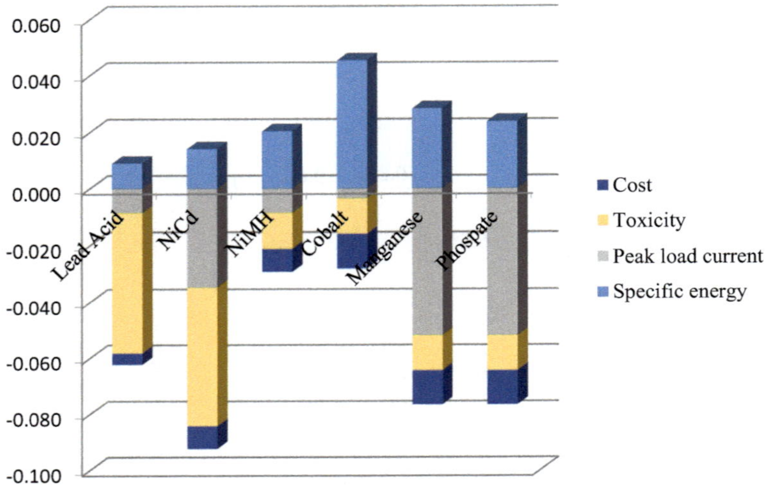

Fig. 4.2 Factor priorities

metal-phosphate brings up the battery efficiency. It is a detailed assessment of the industry utilized in the full mechanism. Technical assessment is a significant principle for power unit mechanism assessment and the battery mechanism's efficiency is a significant principle for its technical assessment. The energy storage power unit's efficiency degree is assessed with regards to the energy storage performance, life-cycling, energy density, and specific capacity.

Life cycling brings up the number of discharge–charge cycles that an energy storage power unit can conduct under a specific capacity. Specific capability and energy intensity bring up the released energy through a power storage battery of unit mass and the contained energy in the power storage battery's unit volume, respectively. Concurrently, regarding the reliability and stability of the mechanism, the battery mechanism's available grade is assessed. Solidity brings up the mechanism capability to apply stable conditions under external effects. Credibility brings up the mechanism capacity to finish defined functions within a defined time and under defined circumstances. The battery mechanisms' technical level has a big effect on the economy and sustainability of the mechanism. The more available the industry and the better the energy storage power unit's efficiency, the higher the energy storage mechanism's sustainability and the stronger the continual process capability. All the same, in the first step of structure, the economy is moderately worse, and the equipment cost may be higher. Modern industry mechanism has a high sustainability.

In this research, information was provided concerning batteries with metal phosphate among commercial batteries. The technical properties of batteries with metal phosphate among commercial batteries were investigated in the markets. Altogether, ten criteria are defined for the assessment of rechargeable batteries. For the evaluation, six rechargeable battery alternatives have been analyzed for consumers. Among these batteries, the choice of the most effective battery was made by AHP technique. Co-based Li-ion batteries were found to provide the best performance by the results obtained with AHP multi-criteria decision-making methodology. The methodologies of this type are discussed and summarized to identify limitations and possibilities. Finally, recommendations and significant results for these implementations are supplied to aid scientists in the progress of their studies.

According to the panel of experts, peak load current, toxicity, and specific energy are ranked as the top three factors determining the attractiveness of the batteries with several criteria also contributing to it. While the factors used in this study highlight Cobalt as the optimum choice, the number of criteria as well as the variety of alternative batteries can be increased in the future. The main contribution of this study is to build a decision-making framework for evaluating the alternatives with the potential to increase the scope with more criteria.

References

1. https://www.cei.washington.edu/education/science-of-solar/battery-technology/
2. B. Huang et al., Recycling of lithium-ion batteries: recent advances and perspectives. J. Power Sources **399**, 274–286 (2018)
3. M.J. Lain, J. Brandon, E. Kendrick, Design strategies for high power vs. high energy lithium ion cells. Batteries **5**(4), 64 (2019). https://doi.org/10.3390/batteries5040064
4. Z. Huang, G. Du, Nickel-based batteries for medium- and large-scale energy storage. Adv. Batteries Medium Large-Scale Energy Storage 73–90 (2015)
5. V. Innocenzi et al., A review of the processes and lab-scale techniques for the treatment of spent rechargeable NiMH batteries. J. Power Sourc. **362**, 202–218 (2017)
6. Memory effect now also found in lithium-ion batteries. Retrieved 5 August 2022
7. S. Koohi-Fayegh, M.A. Rosen, A review of energy storage types, applications, and recent developments. J Energy Storage **2020**(27), 101047 (2019)
8. D.O. Akinyele, R.K. Rayudu, Review of energy storage technologies for sustainable power networks. Sustain. Energy Technol. Assess. (2014)
9. P. Ziemba, Inter-criteria dependencies-based decision support in the sustainable wind energy management. Energies **12**, 749 (2019)
10. J. Watróbski, J. Jankowski, P. Ziemba, A. Karczmarczyk, M. Zioło, Generalised framework for multi-criteria method selection. Omega **86**, 107–124 (2019)
11. I.B. Huang, J. Keisler, I. Linkov, Multi-criteria decision analysis in environmental sciences: ten years of applications and trends. Sci. Total. Environ. **409**(19), 3578–3594 (2011)
12. V. Belton, T. Stewart, *Multiple Criteria Decision Analysis: An Integrated Approach* (Springer, US, 2012)
13. B. Talukder, K.W. Hipel, Review and Selection of Multi-criteria Decision Analysis (MCDA) Technique for Sustainability Assessment, in *BT—Energy Systems Evaluation*, vol. 1 (2021), pp. 145–160
14. L. Li, P. Liu, Z. Li, X. Wang, A multi-objective optimization approach for selection of energy storage systems. Comput. Chem. Eng. **115**, 213–225 (2018)
15. A.G. Olabi, Renewable energy and energy storage systems. Energy **136**, 1–6 (2017)
16. S. Dhundhara, Y.P. Verma, A. Williams, Techno-economic analysis of the lithium-ion and lead-acid battery in microgrid systems. Energy Convers. Manage. **177**, 122–142 (2018)

17. C. Zhang, Y.-L. Wei, P.-F. Cao, M.-C. Lin, Energy storage system: Current studies on batteries and power condition system. Renew. Sustain. Energy Rev. **82**, 3091–3106 (2018)
18. P. Nikolaidis, A. Poullikkas, Cost metrics of electrical energy storage technologies in potential power system operations. Sustain. Energy Technol. Assess. **25**, 43–59 (2018)
19. A. Poullikkas, A comparative overview of large-scale battery systems for electricity storage. Renew. Sustain. Energy Rev. **27**, 778–788 (2013)
20. Y. Yang, S. Bremner, C. Menictas, M. Kay, Battery energy storage system size determination in renewable energy systems: a review. Renew. Sustain. Energy Rev. **91**, 109–125 (2018)
21. https://www.doitpoms.ac.uk/tlplib/batteries/battery_characteristics.php#2
22. N. Imanishi, O. Yamamoto, Rechargeable lithium-air batteries: characteristics and prospects. Mater. Today **17**(1) (2014)
23. https://www.mpoweruk.com/performance.htm
24. J. Garche, C.K. Dyer, P.T. Moseley, Z. Ogumi, D.A.J. Rand, B. Scrosati, Encyclopedia of Electrochemical Power Sources. Newnes 407 (2013). ISBN 978-0-444-52745-5
25. https://www.large-battery.com/news/11.html
26. A. Väyrynen, J. Salminen, Lithium-ion battery production. J. Chem. Thermodyn. **46**, 80–85 (2012). https://doi.org/10.1016/j.jct.2011.09.005
27. D. Linden, T.B. Reddy, (eds.), *Handbook of Batteries*, 3rd edn (McGraw-Hill, New York, 2002). Chapter 35. ISBN 0-07-135978-8
28. *Lithium-Ion Battery Charging Basics* (PowerStream Technologies, 2010). Retrieved 4 December 2010
29. https://www.plarad.de/fileadmin/downloads/FAQ_related%20to%20lithium%20ion%20rechargeable%20battery%20care_DA1_en_0216.pdf
30. D.H.P. Kang, M. Chen, O.A. Ogunseitan, Potential environmental and human health impacts of rechargeable lithium batteries in electronic waste. Environ. Sci. Technol. **47**(10), 5495–5503 (2013). https://doi.org/10.1021/es400614y
31. https://batteryuniversity.com/article/bu-808c-coulombic-and-energy-efficiency-with-the-battery
32. P. Kurzweil, Gaston Planté and his invention of the lead–acid battery—the genesis of the first practical rechargeable battery. J. Power. Sourc. **195**(14), 4424–4434 (2010)
33. P. Lyu, X. Liu, J. Qu, J. Zhao, Y. Huo, Z. Qu, Z. Rao, Recent advances of thermal safety of lithium-ion battery for energy storage. Energy Storage Mater. **31**, 195–220 (2020)
34. X. Liu, Y. Li, X. Xu, L. Zhou, L. Mai, Rechargeable metal (Li, Na, Mg, Al)-sulfur batteries: materials and advances. J. Energy Chem. **61**, 104–134 (2021)
35. T. Oshima, M. Kajita, A. Okuno, Development of sodium-sulfur batteries. Int. J. Appl. Ceram. Technol. **1**(3), 269–276 (2005)
36. Y. Yang, H. Yang, X. Wang, Y. Bai, C. Wu, Multivalent metal–sulfur batteries for green and cost-effective energy storage: current status and challenges. J. Energy Chem. **64**, 144–165 (2022)
37. M. Assefi et al., Recycling of Ni-Cd batteries by selective isolation and hydrothermal synthesis of porous NiO nano cuboid. J. Environ. Chem. Eng. **6**(4), 4671–4675 (2018)
38. D. Bresser, E. Paillard, S. Passerini, Lithium-ion batteries (LIBs) for medium- and large-scale energy storage. Adv. Batter. Medium Large-Scale Energy Storage 125–211 (2015)
39. A. Perner, J. Vetter, Lithium-ion batteries for hybrid electric vehicles and battery electric vehicles. Adv. Battery Technol. Electric. Veh. 173–190 (2015)
40. M. Morris, S. Tosunoglu, Comparıson of rechargeable battery technologies, in *ASME Early Career Technical Journal 2012 ASME Early Career Technical Conference* (ASME ECT, Atlanta, Georgia USA)
41. E. Mu, M. Pereyra-Rojas, *Practical Decision Making: An Introduction to the Analytic Hierarchy Process (AHP) Using Super Decisions*, vol. 2 (Springer, 2016)

Chapter 5
Impact of Phase Change Material on Greenhouse Energy Balance Under Light Abatement Curtains

Quade Digweed and Sabrina Sawan

Abstract The influence of long photoperiod lighting, light abatement curtains and phase change materials on greenhouse energy balance was investigated in two independent greenhouses at the Harrow Research and Development Centre. Two identical greenhouses (50 m^2 area) were outfitted with Svensson 9950 light abatement curtains, twelve modified 600W Lumi-Gro top lights, and 80 Flatice 21 °C Phase Change Material Ice packs, with total thermal storage capacity of 15.6 kWh. Both greenhouses were planted with cluster tomatoes cv. 'Endeavor' in December 2022 and grown until May 2023. The greenhouse microclimate, including air temperature, humidity, crop canopy temperature, lighting, and greenhouse surface temperatures were monitored throughout the experiment. From December to March, one greenhouse was subjected to 24 h continuous lighting, while the other was placed under a conventional 16 h lighting strategy (02:00–18:00). In both houses, curtains were fully closed from dusk until dawn. The phase change material was found to be incompatible with a 24 h lighting strategy due to the lack of temperature fluctuations for full charge/discharge cycles. In March 2023, the phase change material was removed from one house and both houses were switched to 16 h lighting strategies to observe the impact of phase change material on greenhouse energy balances. The phase change material was able to meet 5 h of heating demand when compared to the control treatment. No difference in yield was found between trials. This case study indicates that phase change material can augment traditional heating systems to reduce the need for natural gas while maintaining yield.

Q. Digweed (✉) · S. Sawan
Harrow Research and Development Centre, Agriculture and Agri-Food Canada, 2585 County Rd. 20, Harrow, ON N0R 1G0, Canada
e-mail: quade.digweed@agr.gc.ca

D. S-K. Ting and A. Vasel-Be-Hagh (eds.), *Engineering to Adapt*, Springer Proceedings in Energy, https://doi.org/10.1007/978-3-031-47237-4_5

5.1 Introduction

Greenhouses have been used to grow horticultural crops achieving high productivity in Canada for decades, with many growers now aiming for year-round production practices. To meet the demand for year-round production, the use of supplemental lighting in Ontario vegetable greenhouses has accelerated rapidly, with an estimated 282% increase in electricity usage from 2018 to 2024 [1]. Most of the growth in lighting use has occurred in the Essex region of Ontario. This high concentration of greenhouses using supplemental lights has caused visible light pollution in the night sky and measurable nighttime light emissions [2]. Similar nighttime greenhouse light emissions occur in other regions with greenhouse density, such as Niagara. To address this issue, municipal governments across Canada have passed light abatement bylaws, mandating the use of light abatement curtains at night in greenhouses using supplemental lighting. In addition to preventing stray light emissions, light abatement curtains increase greenhouse insulation and limit dehumidification by venting due to reduced moisture transport through the curtain material. The closure of curtains while supplemental lights are in use can cause increased greenhouse temperatures and humidity, with potentially adverse effects on greenhouse crops. Simultaneously, there is an opportunity to use light abatement curtains to reduce greenhouse heating demand and harvest waste heat produced from light fixtures. Modeling work by Katzin et al. comparing light emitting diode (LED) and high pressure sodium (HPS) lighting technologies has shown that this waste heat from light fixtures can reduce, or even eliminate, supplemental heating demand in greenhouse production systems [3, 4].

A key challenge to direct heating via light fixtures in greenhouse production is that peak heating demand coincides with nighttime, when plants are allowed to rest, although some practices, such as temperature dips, can decrease the minimum length of this rest period [5]. Additionally, increasing light levels under light abatement curtains to meet heating demand will also increase the crop transpiration rate, leading to higher humidity levels, compounding challenges with humidity transfer through curtains. The use of light abatement curtains creates two major challenges to be addressed: overheating due to trapped waste heat from greenhouse lights, and excessive humidity that could lead to disease development and poor crop transpiration rates.

Previous experiments conducted at the Harrow Research and Development Centre (RDC) have demonstrated that the greenhouses available for this experiment cannot maintain elevated humidity rates, even under light abatement conditions, due to high surface area to volume ratios providing ample greenhouse exterior surfaces for condensation of water vapor [6]. For this reason, the study conducted in the winter of 2022–2023 was focused on management of heat under light abatement curtains. Identified technologies for consideration in the trial were phase change materials, long photoperiod lighting, and hot water buffer tank storage.

5.1.1 Phase Change Material

Phase change materials (PCM) are substances that undergo a phase change (generally solid–liquid) at a desired process temperature. The significant latent heat storage or release at this process temperature can help maintain a desired setpoint in a greenhouse and has been demonstrated in solar greenhouses to act as a heat sink when incorporated in the north wall [7, 8]. A PCM bank could be installed in a greenhouse to capture waste heat from greenhouse lights for release during the dark period, simultaneously reducing the magnitude of overheating under light abatement, and reducing energy usage for supplemental heat during the dark period.

Commercially available PCMs which offer phase change temperatures near greenhouse setpoints consisted mainly of salt hydrates and organic paraffin and non-paraffin, with a range of latent heat storage capacities [9]. Focusing on the preferred greenhouse setpoints between 18 and 23 °C, salt hydrate PCMs had slightly higher storage capacities than organic PCMs [10]. Furthermore, salt hydrate PCMs are available in encapsulated bricks easily interlocked for location under greenhouse troughs to minimize space and avoid shading of the crop. A salt hydrate PCM, S21 (PCM Products, Cambridgeshire, UK), was selected for further greenhouse testing in this study.

5.1.2 Long Photoperiod Lighting

Traditional greenhouse lighting strategies provide the crop with up to 16 h of supplemental light, until a target daily light integral (DLI) is met for the day. This lighting period is usually started after midnight and continues until natural sunset (18:00). It is generally accepted that exceeding this period can cause injury to greenhouse crops [5]. Recent developments in long photoperiod lighting by Lanoue et al. have demonstrated that longer photoperiods, up to 24 h, are possible with dynamic lighting strategies that alternate the use of red and blue light spectra [11]. Such lighting strategies deliver the same target DLI as 16-h strategies but can reduce the light intensity required by up to 33%. The extension of the lighting period to 24 h would reduce the need for supplemental heating during the night due to waste heat from light fixtures, and the reduction in intensity would similarly reduce the overheating effect. Because this strategy had already been demonstrated on tomatoes, it was included for consideration in the study.

5.1.3 Hot Water Storage Tanks

Hot water storage tanks are a well-developed technology used in commercial greenhouses in Canada. In these systems, a large volume insulated tank is connected to the greenhouse hot water heating system. This tank allows for the decoupling of boiler operation from heating demand. The boiler can be run during the day to produce CO_2 that is captured, scrubbed from flue gases, and then supplemented into the greenhouse while the plants are photosynthesizing and require CO_2. The hot water produced by the boiler is pumped into the storage tank, and when heating is required at night, the hot water stored in the tank can be used before the boiler is again called on to meet the heating demand. Past studies have considered the addition of PCMs to the tanks to increase storage capacity, with favorable results [12]. While further study of hot water storage tanks in combination with PCM was considered, it was not possible to subdivide the greenhouse heating system at the Harrow RDC for this purpose.

As a result of the above considerations, this study chose to focus on the application of PCMs directly in the greenhouse environment, rather than in a separate storage system or hot water buffer tank, and in combination with long photoperiod lighting strategies to determine their compatibility.

5.2 Methodology

Greenhouse tomato cv. 'Endeavor' was chosen as the crop to use in the study. It is the most commonly grown greenhouse crop in Ontario [13], and the long photoperiod lighting recipe has been previously tested on this cultivar. The crop was planted on December 13th, 2022 and grown until May 2nd, 2023.

5.2.1 Greenhouse Properties and Equipment

The two greenhouses used in the study are 7.2 m × 6.5 m in area, with a gutter height of 3.3 m. Both houses were identical, with double-layer polyethylene cladding, east facing ridge vents the length of the greenhouse, insect screens, light abatement curtains, and supplemental LED lighting. Heat was provided from four ground level hot water heating rails with total surface area of 10.46 m^2 and water flow rate of 1.1 L/s. Both houses were controlled by a Priva climate computer. A summary of equipment installed for the study is provided in Table 5.1.

The installed light abatement curtain claims to prevent 99% of light emissions while conserving 50% of the greenhouse heat. The installed supplement lighting

Table 5.1 Mini greenhouse equipment

Equipment	Specifications	Quantity
Light abatement curtains	Svensson Obscura 9950 FR [14]	1
Supplemental lighting	LumiGrow Pro 650e [15]	12
Phase change material	PCM Products Plusice S21 Flatice [10]	78

Table 5.2 Greenhouse climate settings

Time	Heating setpoint (°C)	Cooling setpoint (°C)	Humidity setpoint (%)
2:00–17:00	21	26	75
17:01–1:59	19	22	75

Table 5.3 Greenhouse lighting schedule

Treatment	Period 1	Intensity (umol/ m^2/s)	Period 2	Intensity (umol/ m^2/s)	Total DLI (mol/ m^2/d)
16 h	2:00 to 18:00	250	18:01 to 1:59	0	14.4
24 h	2:00 to 18:00	200	18:01 to 1:59	100	14.4

fixtures consume 650 W of electricity at full power and have spectrum and intensity controls. The fixture controls allow for dimming down to 20% of full power, and independent control of red, blue and white light channels. The 78 PCM packs per greenhouse contain a salt hydrate solution with a melting temperature of 21 °C, and provided a total thermal storage capacity of 15.6 kWh The greenhouse climate settings are listed in Table 5.2.

Cooling and humidity control was realized by roof vent opening. To aid in air movement, a minimum pipe temperature of 30 °C was set. Light abatement curtains were deployed at 17:00 and opened at 7:00. Greenhouse lighting schedules are shown in Table 5.3.

The spectral intensity of both light treatments are shown in Fig. 5.1. The 16-h photoperiod and 24-h day period were 40% blue, 11% white and 49% red. The night period for the 24-h treatment was 93% red. The target was 100% red but there was a measured glow from the blue LED chips when digitally turned off that can be seen in the Fig. 1b. Measurements were taken with a Li-180 spectrometer (Li-COR Biosciences. Lincoln, Nebraska) under light abatement curtains at 3 locations per treatment.

5.2.2 *Tomato Crop Management*

120 stems per greenhouse of cluster tomatoes cv. 'Endeavor' were planted on December 13th, 2022 at a density of 2.56 stems/m^2 in 4 rows of 30 stems each in rockwool on raised gutters. The two outer rows were subject to edge effects and not used for data collection. Plants were fertigated with a standard greenhouse tomato solution recommended in the Ontario greenhouse vegetable guide [16]. Tomatoes were harvested weekly when a cluster was 60% ripe. Cluster weight was recorded from the middle two rows. Harvests continued until May 2nd, 2023, when remaining tomatoes were harvested and green tomatoes counted and recorded separately as potential yield.

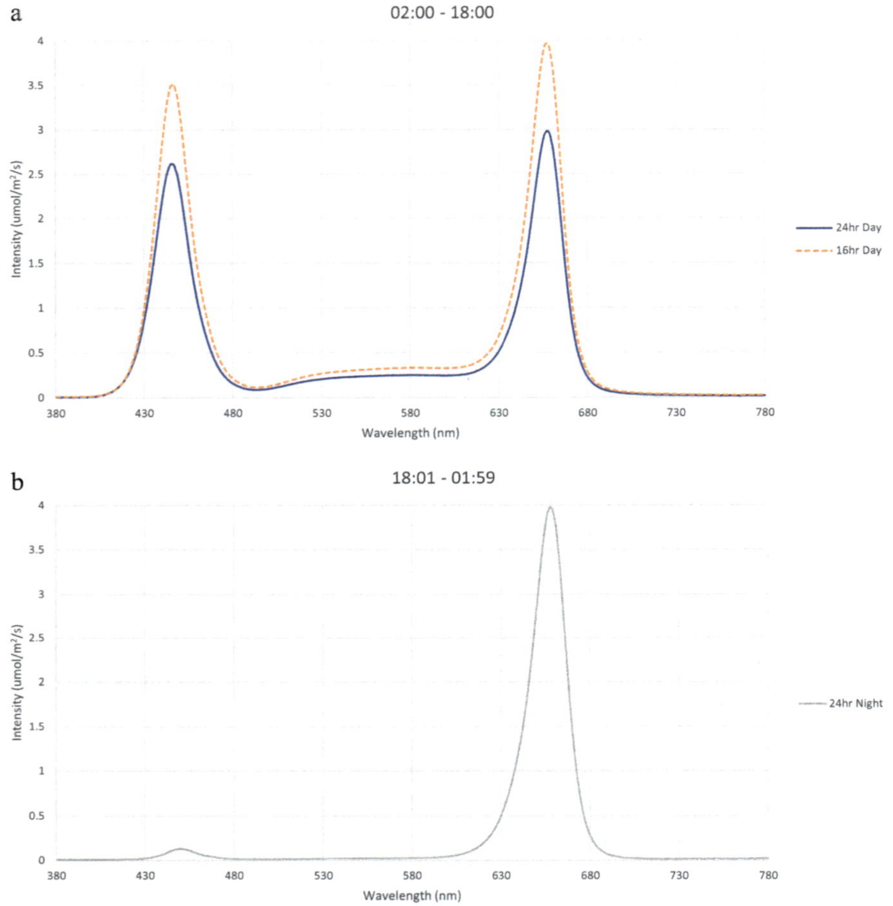

Fig. 5.1 Intensity by wavelength of supplemental lighting treatments. **a** Day treatment, **b** night treatment

5.2.3 Environmental Data Collection

The greenhouse climate was monitored using a Campbell Scientific CR1000X [17] datalogger in each greenhouse with two AM16/32 multiplexers and an AM25T Thermocouple multiplexer. Surface temperatures were recorded using type T thermocouples fixed to the surface with reflective tape. Captured surface temperatures include the greenhouse walls, floor, roof, phase change material and heat pipes. Crop canopy temperature was captured using Omega Engineering OS211 LT IR transmitters [18], and air temperature and relative humidity were collected with analog NTC thermistors [19] and Honeywell HIH 4000 sensors [20]. Canopy temperature, air temperature, and humidity was collected at 1 m above the greenhouse gutter and at the plant head. Data was collected every minute and the 5 min average was recorded to the CR1000X's internal memory.

5.2.4 Calculation of Heating Energy Use

The greenhouses used for the trial have perimeter and rail heating systems. Because the compartments were retrofitted with blackout curtains, the perimeter heating system was behind the interior curtains and did not directly contribute to heating the greenhouse crop. For the experiment, the perimeter heat was turned down to only provide heat to melt any frost forming on the greenhouse cover materials, and the rail heat was relied upon as the primary greenhouse heating method.

Greenhouse heat is provided from main trunk lines back to the greenhouse boiler, with mixing valves leading to each greenhouse. The mixing valve position are controlled by the Priva climate computer to meet the greenhouse heating setpoint, and water is constantly circulated at the 1.1 L/s flow rate in the greenhouse compartments. Because of the high flow rate relative to the pipe length, pipe inlet and outlet temperatures in the house are nearly identical and heat input can't be calculated from the flow rate and temperature difference. Instead, radiant and convective heat transfer was assumed to occur, and heat transfer from the pipe surface to the environment was calculated in each house.

The effect of solar radiation heating the pipes also had to be considered. Often, greenhouse surface temperatures would exceed greenhouse air temperature under high solar radiation, which would lead to heat transfer caused by solar radiation to be attributed to heat input from the boiler. To avoid this, any heat transfer from the pipes was not counted to energy input when the greenhouse air temperature was above the heating setpoint, as this heat would have originated from solar radiation and not from the boiler.

5.2.4.1 Convective Heat Transfer

Convective heat transfer from the greenhouse pipes to the greenhouse air was calculated using Eq. (5.1) below:

$$Q_{conv} = hA_{pipe}\left(T_{pipe} - T_{air}\right) \tag{5.1}$$

With h being the convective heat transfer coefficient (taken to be 8 W/m^{2}K), A_{pipe} is the rail heating pipe surface area, T_{pipe} is the pipe temperature, and T_{air} is the air temperature. Heat flux was calculated in 5 min intervals.

5.2.4.2 Radiant Heat Transfer

Radiant heat transfer from the greenhouse pipes into the greenhouse environment was calculated using Eq. (5.2) below:

$$Q_{Rad} = \varepsilon \sigma A_{pipe} \left(T_{pipe}^4 - T_{surr}^4 \right) \quad (5.2)$$

With T_{surr} being the average temperature of surrounding objects and greenhouse air, ε being the pipe emissivity (taken to be 0.95), and σ the Stefan-Boltzmann constant. The average surrounding temperature was estimated using view factors calculated using the Monte-Carlo and analytical determination from Kaname Sasaki's view factor calculator [21]. The sum of all surface temperatures, multiplied by their view factor, was used to calculate an average surrounding temperature. The position of the greenhouse pipe relative to crops, PCM, ground, and roof was taken as shown in Fig. 5.2.

Table 5.4 describes the surfaces assumed to absorb radiation from the heat pipe. Each surface listed in Table 5.4 was assumed to be a rectangular surface 90 degrees to a differential surface of pipe, subject to Eq. 5.3.

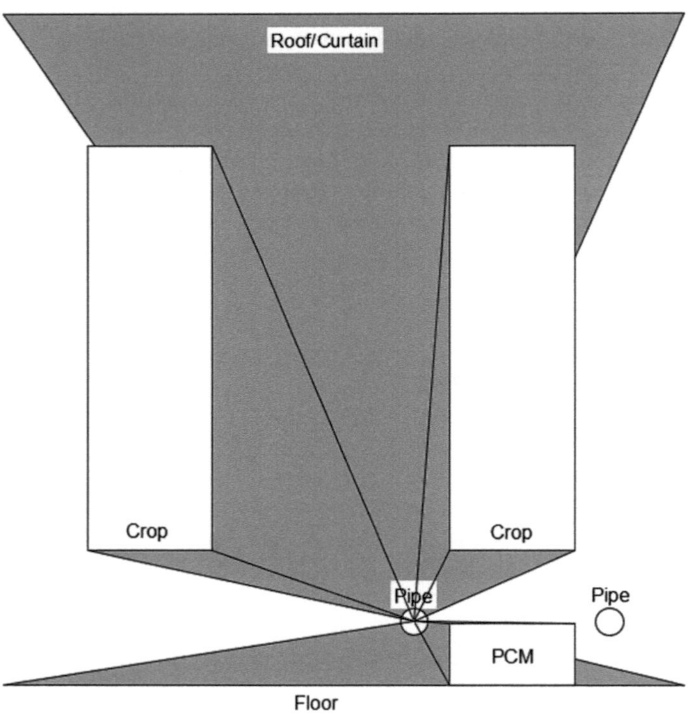

Fig. 5.2 Projected view factors of heating pipe in simplified greenhouse model

Table 5.4 Rectangular surface approximations of greenhouse objects absorbing radiation from the heating pipe

Object	Length (a)	Width (b)	Distance to pipe (c)	View factor
Greenhouse floor	6.5	7.2	0.17	0.24
Crop row above pipe	2.08	5.2	0.76	0.16
Crop row adjacent	2.08	5.2	1.8	0.08
Greenhouse roof/curtain	5.2	6.2	2.3	0.14
Phase change material	0.23	5.1	0.1	0.15

$$F = \frac{1}{2\pi}\left(\arctan\frac{1}{Y} - \frac{Y}{\sqrt{X^2 + Y^2}}\arctan\frac{1}{\sqrt{X^2 + Y^2}}\right) \tag{5.3}$$

where X = a/b, and Y = c/b. To reach unity, the balance of radiation was assumed to reach the greenhouse walls as seen by the empty space in Fig. 5.2.

5.2.5 *Phase Change Material Experiment*

Midway through the month of March, it was observed that the PCM in the 24 h lighting trial was not melting or solidifying completely, and with the current setup, a 24-h lighting strategy was not compatible with PCM used as an energy storage system. At this time, the phase change material was removed from one greenhouse and both greenhouses switched to the 16-h settings. A sub-experiment testing only the efficacy of phase change material was conducted from March 31st to May 2nd.

5.3 Results and Discussion

In both greenhouses, a successful tomato crop was grown. Harvest began on March 10th and continued for 9 weeks. There were minimal differences in yield at the end of March 2023, when the 24 h lighting strategy was in use in one house. Comparative yields can be seen in Fig. 3a. Yield before week 7 can be attributed to the light experiment. There was observed chlorosis from the 24-h lighting strategy on upper leaves of the plant canopy but leaf injury localized to regions of higher light intensity due to the fixture proximity to the plants.

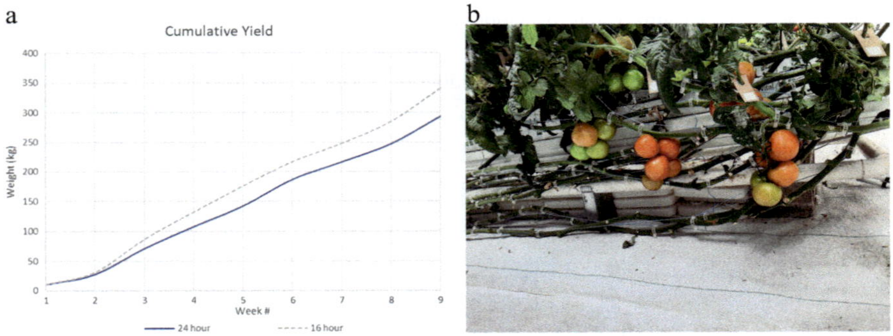

Fig. 5.3 **a** Cumulative yield under light treatments, and **b** tomatoes ripening for harvest in PCM treatment

5.3.1 Combined PCM and Light Treatments

When the use of PCM was combined with 24-h lighting treatments, there was minimal difference between energy usage in the two compartments as seen in Fig. 4a. Figure 4b shows that the PCM in the 24-h light treatment never underwent complete phase change, and stayed solid through the day, while the 16-h light treatment allowed the phase change to freeze at night and melt during the day. Physical observations of the 24-h treatment PCM showed that the phase change packs were a slurry and only melted on the exterior of the pack, not all the way through. This indicates that the rate of heat transfer in and out of the packs was not high enough to fully melt a pack in a 24-h period. For the following phase, a thermocouple was fixed between two lower packs, rather than two upper packs, to measure the internal temperature.

5.3.2 PCM Experiment

From March 31st to May 2nd, the 24-h light treatment was switched to a 16-h light treatment and phase change material was removed from the greenhouse to study the effect on the greenhouse climate and energy use without the influence of the lighting experiment. Figure 5a shows the cumulative heat energy used over this time period, while Fig. 5b shows the phase change temperature vs. the floor temperature in the greenhouse with no PCM.

In Fig. 5a, the daily heat flux through the month of April in the control greenhouse (i.e., no PCM) can be seen to remain constant, despite increasing outdoor temperatures and high daily greenhouse temperatures that can be seen in Fig. 5b.

Figure 6a and b show the relationship between greenhouse air temperature and venting percentages averaged over the month of April 2023. Figure 6c and d show the relationship between PCM temperature and heating pipe temperature averaged over the month of April 2023. Figure 6a shows that the average air temperatures between

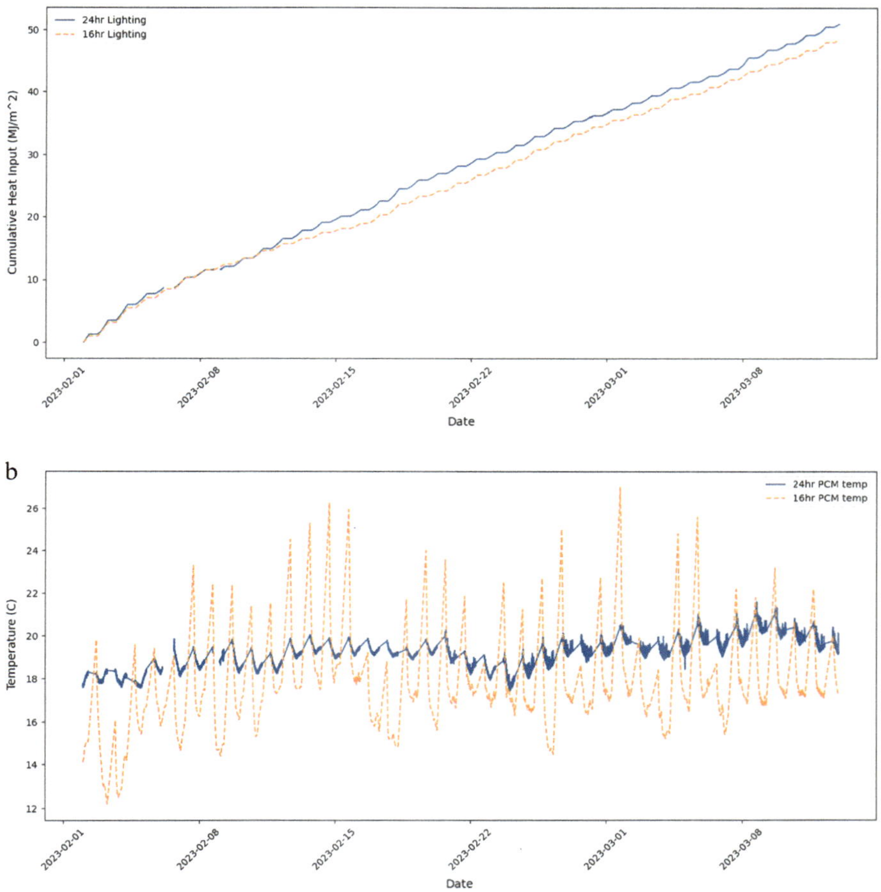

Fig. 5.4 **a** Heating energy input by lighting strategy and **b** PCM temperature by lighting strategy

treatments was relatively constant. The PCM successfully stored heat during the day, keeping the average greenhouse temperature a degree lower, but was unable to cool rapidly when switching to a night temperature. This cooling attempt is most visible in graph 6b, where the PCM treatment vent is open to try and reach the lower nighttime setpoint, causing the PCM to discharge (see Fig. 6c) before the nighttime setpoint is reached. Looking to Fig. 6c, while the then fully discharged PCM begins to charge at 2:00, when the supplemental lighting turns on, no change in temperature is observed until the following dawn, and the PCM is fully charged by 9:36, when it passes 21 °C from solar radiation. Figure 6d shows the reduction in heating demand in the PCM treatment, with the exception of early morning between 7:00 and 9:30. When the greenhouse curtain opens at 7:00, the pipe temperature is elevated in both houses to heat the cooler attic air reaching the greenhouse crop. Because of the PCM was located adjacent to the heat pipes, a portion of the heat is diverted to storage instead

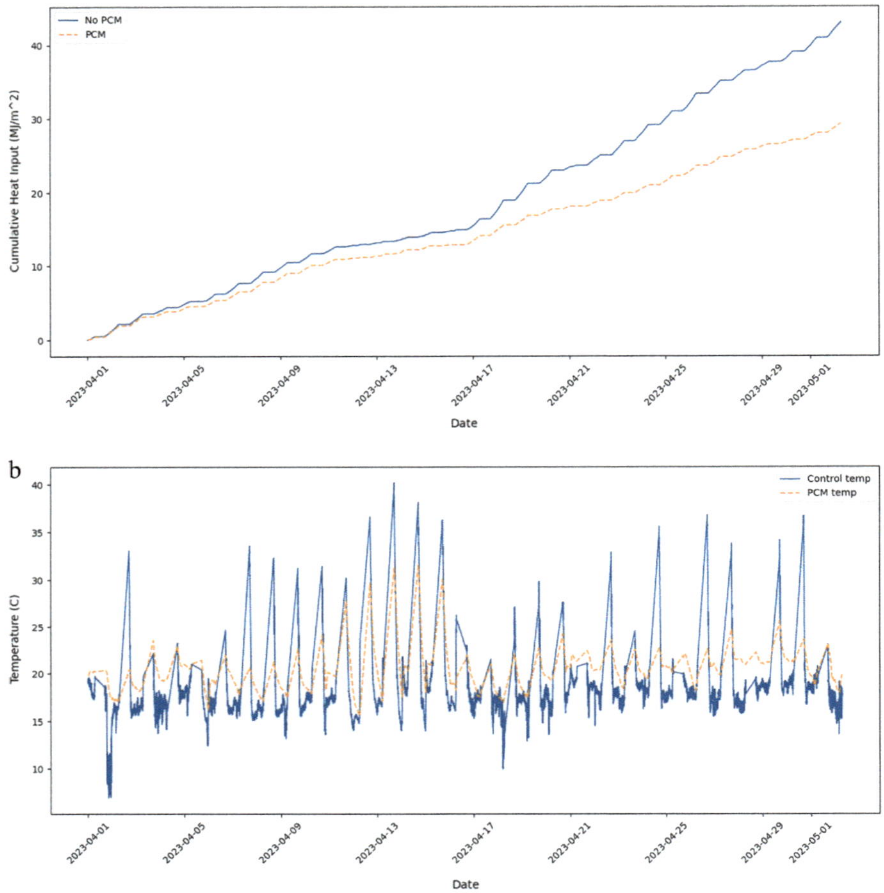

Fig. 5.5 **a** Heating energy input by treatment and **b** PCM temperature compared to control greenhouse floor temperature

of heating the greenhouse space in the PCM treatment until the PCM is melted. This charging of the PCM with pipe heat, rather than solar radiation, is undesirable. It diminishes the PCM ability to maintain cooler greenhouse temperatures through the day and lowers overall PCM efficiency. Therefore, subsequent experiments should aim to optimize PCM placement within the greenhouse environment.

Figure 5.7 below shows a kernel density estimation of the heating input required at a given PCM or control temperature, with frequency distributions of observed temperatures. Overall, lower temperatures in the PCM system vs. the control correlated with lower heating requirements, as would be expected. The PCM system also demonstrates lower variance in heating requirements due to the added thermal mass of the system. The x axis of Fig. 5.7 shows the univariate kernel density estimation

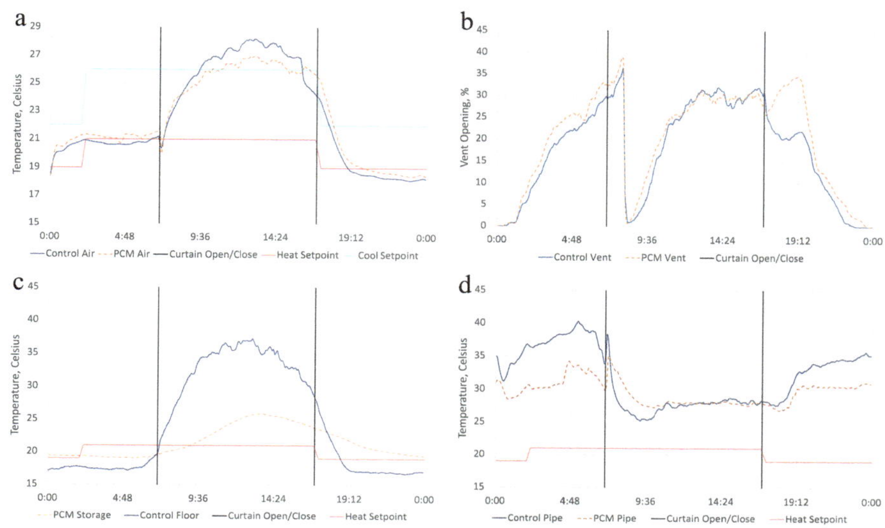

Fig. 5.6 Daily climate averages for the month of April. **a** Average greenhouse air temperature, **b** average greenhouse vent opening, **c** average PCM and floor temperature, and **d** average pipe temperature

of temperature. The y axis of Fig. 5.7 shows the univariate kernel density estimation of heat input, demonstrating the lower and more stable heat input in the PCM greenhouse.

5.4 Summary of Heating Demands

Table 5.5 summarizes the difference in heating demand between treatments. While there was a slight increase in heating demand for the long photoperiod treatment in combination with PCM, it was minimal. When PCM storage was used with a conventional photoperiod during the spring season, a 31.9% decrease in total heating demand was observed.

The results presented in Table 5.5 also demonstrate efficiencies that could have been realized in greenhouse setpoints. The heating demand for April in the control was nearly as high as February, despite the change in outdoor temperatures. This can be attributed to the minimum pipe temperature set to 30 °C, and the use of vents for humidity control. Looking to Fig. 6b and d, the vents can be seen open while the pipe temperature is elevated. This indicates that the compartments were venting due to humidity, even though additional heat was required (Fig. 5.8). This need for increased ventilation under light abatement curtains should be the focus of future work, otherwise heat saving benefits of the curtains will be offset by current dehumidification strategies.

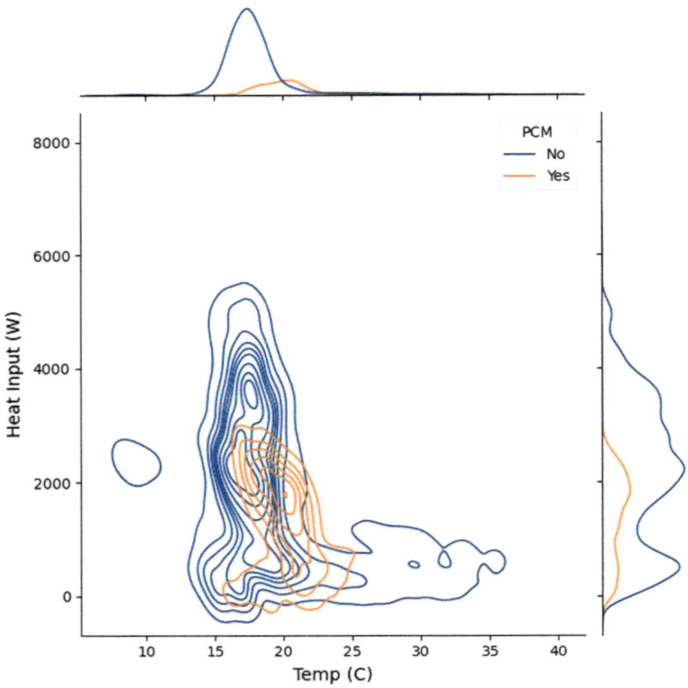

Fig. 5.7 Bivariate kernel density estimation and distribution of heating demand as predicted by PCM or floor temperature. Univariate kernel density estimation of Temperature and Heat Input shown on their respective axes

Table 5.5 Heating demand summary

	Heating Energy Use (MJ/m^2)	Percent difference
Lighting experiment		
Control (16 h)	48.5	–
Long photoperiod (24 h)	51.0	+5.09%
PCM experiment		
Control	43.0	–
PCM storage	29.3	–31.9%

5.4.1 PCM Location and Control

The increased venting at 17:00 and increased heating at 7:00 in Fig. 6a–d demonstrates the need for increased PCM control for full integration into current greenhouse production systems. While locating the PCM under the greenhouse gutter causes the

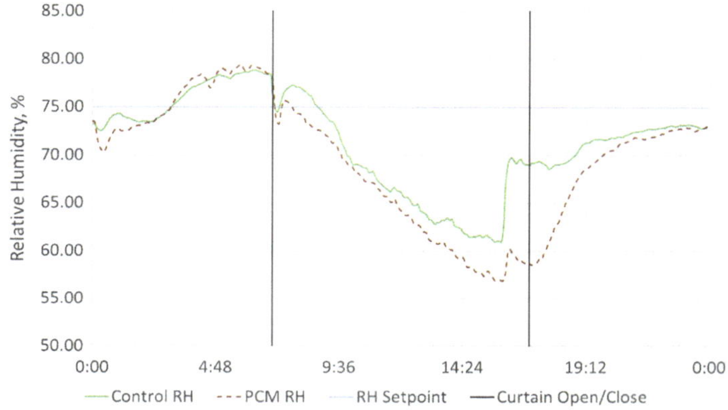

Fig. 5.8 Relative humidity in control and PCM greenhouse compartments

least interference with conventional greenhouse maintenance, it has several disadvantages. The PCM is highly shaded from solar radiation in this location, minimizing the ability to store this energy for later use, and is instead charging from pipe heat or convection from the greenhouse air. Additionally, the under gutter location causes the PCM to have a high view factor when considering pipe radiation, reducing the efficacy of pipe heating in the early morning when the PCM is discharged.

One of the major disadvantages of the PCM is the inability to control the charge/discharge cycle. Greenhouse systems rely on dynamic temperature control through the day, and PCM directly counters that dynamic control by attempting to maintain a static temperature. Furthermore, existing greenhouse control systems are not tuned for the inclusion of PCM, so expected heating and cooling behaviour are negatively impacted, as seen in Fig. 6b and d, even though the overall heat usage is lower. Direct incorporation of the PCM into the greenhouse control system would be recommended for future work. This could be achieved in several ways. First, the PCM could be added to the existing hot water storage tanks attached to commercial greenhouses, to extend the capacity of the tank. This is discussed in the Naghibi et al. paper [12] but could not be tested in this experiment due to greenhouse design.

Secondly, the PCM could be added to a standalone insulated tank and ducted into the greenhouse system, to rely on air-PCM convective heat exchange. In this system, the current ideal location underneath the gutter could still be used, but the effects of pipe heating on the PCM could be minimized with aluminum reflective coverings on the array. Controls could be added to open a louver and activate a fan to force air through the PCM system. Once refined, a variable frequency drive could be added to the fan to precisely control the charge/discharge rate of the PCM tank as required.

Thirdly, the PCM could be relocated in the greenhouse to locations with higher solar radiation that do not impact incident solar radiation to the crop, such as PCM tiles on the greenhouse floor, greenhouse sidewalls, and inside of greenhouse structure. This solution has the advantage of low integration requirements with greenhouse controls but would require a redesign of the current greenhouse structure, and retrofitting would not be possible beyond adding tiles to the greenhouse floor. This greenhouse floor option would have a similarly high view factor to the heating pipes as the current solution, with the associated drawbacks.

5.5 Conclusions

The experiment conducted at Harrow RDC through the winter of 2022–2023 examined the impact of both long photoperiod lighting and PCMs on greenhouse energy balance. Several conclusions can be drawn, along with several needs for future work.

5.5.1 *Outcomes of Experiments*

1. Current PCM strategies added passively to greenhouses are not compatible with 24-h lighting strategies.
2. PCMs reduce greenhouse energy usage during shoulder seasons, where there are warm daytime temperatures to fully charge the PCM and cool nights to fully discharge the PCM. Their efficacy is limited in current greenhouse systems due to their tendency to capture pipe heat before solar radiation if used to meet nighttime heating demand.
3. The energy-saving potential of light abatement curtains is limited if the high humidity environment under the curtain causes the vents to open. The greenhouse pipe temperature will rise to meet heating setpoints while venting. Alternative dehumidification strategies are required to optimize curtain use.

5.5.2 *Future Work*

1. The current PCM system was passive. To reduce impact of the PCMs on the pipe heating system's efficiency, the PCM system needs to be incorporated into the greenhouse controls. This could be done with a new system, or by incorporation into a greenhouse hot water tank.
2. Dehumidification technologies designed to work with PCMs and greenhouse lighting to meet dehumidification targets without relying on heating and venting will increase curtain effectiveness. Such a new system needs to be incorporated into the curtain management and thermal storage system.

3. Optimized PCM location within the greenhouse to reduce the syphoning of heating energy from the heat pipes is of interest for future studies.

Acknowledgements The authors would like to acknowledge the Ontario Greenhouse Vegetable Growers for funding this work through the Canadian Agricultural Partnership, and the University of Guelph for providing the light abatement curtains used in one of the greenhouses. Greenhouse setup and lighting layout was possible with the help of Drs. Xiuming Hao and Jason Lanoue at the Harrow RDC. Thanks to the Harrow RDC greenhouse crew for the crop maintenance through the experiments, and to Matthew Prior and Alex Nauta for their help in modifying the light fixtures used in this experiment.

References

1. Posterity Group, Increasing Electricity Use in Ontario's Greenhouse Sector (2019). https://www.ieso.ca/-/media/Files/IESO/Document-Library/research/Greenhouse-Infographic.ashx. Accessed 14 June 2023
2. K. Walczak, A Land of Perpetual False Dawn (2021). https://www.darksky.org/light-pollution-industrial-greenhouses/. Accessed 14 June 2023
3. D. Katzin, L. Marcelis, S. van Mourik, Energy savings in greenhouses by transition from high-pressure sodium to LED lighting. Appl. Energy **281**, 116019 (2021). ISSN 0306-2619. https://doi.org/10.1016/j.apenergy.2020.116019
4. D. Katzin, L. Marcelis, E. van Henten, S. van Mourik, Heating greenhouses by light: a novel concept for intensive greenhouse production. Biosyst. Eng. **230**, 242–276 (2023). ISSN 1537-5110. https://doi.org/10.1016/j.biosystemseng.2023.04.003
5. X. Hao, X. Guo, X. Chen, J. Zheng, L. Celeste, S. Khosla, Dynamic temperature integration with temperature drop improved the response of greenhouse tomato to long photoperiod of supplemental lighting. Acta Hortic. **1170**, 995–1002 (2017). https://doi.org/10.17660/ActaHortic.2017.1170.128
6. Q. Digweed, Lighting and Abatement, Proceedings of Canadian Greenhouse Conference, 5 October 2022. https://drive.google.com/drive/folders/1NUPOGtmND4si62dqG_2HZKDvKZDOnBfw
7. L. Peng, Z. Ya-hong, B. Qing, H. Wei, J. Li, Z. Xue-ning, Analysis on thermal characteristics of trapezoidal wall based on phase change materials in solar greenhouse. Chin. J. Agrometeorol. https://doi.org/10.3969/j.issn.1000-6362.2019.10.002
8. F. Berroug, E.K. Lakhal, M. El Omari, M. Faraji, H. El Qarnia, Thermal performance of a greenhouse with a phase change material north wall. Energy Build. **43**(11), 3027–3035 (2011). ISSN 0378-7788. https://doi.org/10.1016/j.enbuild.2011.07.020
9. E. Paroutoglou, A. Afshari, N. Bergsøe, P. Fojan, G. Hultmark, A PCM based cooling system for office buildings: a state-of-the-art review, in *E3S Web of Conferences* (2019). https://doi.org/10.1051/e3sconf/201911101026
10. PCM Products, PlusIce™ product range. https://www.pcmproducts.net/files/PlusICE%20Range%202021-1.pdf. Accessed 14 June 2023
11. J. Lanoue, J. Zheng, C. Little, A. Thibodeau, B. Grodzinski, X. Hao, Alternating red and blue light-emitting diodes allows for injury-free tomato production with continuous lighting. Front. Plant Sci. **10** (2019). https://doi.org/10.3389/fpls.2019.01114
12. Z. Naghibi, R. Carriveau, D.S.-K. Ting, Improving clean energy greenhouse heating with solar thermal energy storage and phase change materials. Energy Storage **2**, e116 (2020). https://doi.org/10.1002/est2.116

13. Ontario Greenhouse Vegetable Growers, OGVG Fact Sheet. https://www.ogvg.com/_files/ugd/5ef796_985323da14254ea5801696693d935ad8.pdf. Accessed 15 June 2023
14. Svensson, Obscura 9950 FR W Specifications. https://www.ludvigsvensson.com/en/climate-screens/climate-screens-products/obscura/obscura-9950-fr-w. Accessed 15 June 2023
15. LumiGrow, Pro Series E Smart Horticultural Lighting, Specifications Sheet PN #770–00003-K (2017). https://cdn.thomasnet.com/ccp/30842909/264677.pdf
16. Ontario Ministry of Agriculture, *Food and Rural Affairs, Growing Greenhouse Vegetables in Ontario* (King's Printer for Ontario, Publication 836, 2010)
17. Campbell Scientific, CR1000X Measurement and Control Datalogger. https://www.campbellsci.ca/cr1000x. Accessed 19 June 2023
18. Omega Engineering, Fixed Infrared Temperature Transmitters with a Selection of Field of Views to fit most Industrial Applications. https://www.omega.ca/en/temperature-measurement/noncontact-temperature-measurement/fixed-infrared-temperature-sensors/p/OS210-150-300-800. Accessed 19 June 2023
19. Amphenol, Epoxy type NKI noise immune thermistor. 10 000 type 5 NKI#103C5*1^. https://www.amphenol-sensors.com/en/thermometrics/ntc-thermistors/epoxy/3447-type-nki. Accessed 19 June 2023
20. Honeywell, HIH-4000 Series. https://sps.honeywell.com/ca/en/products/advanced-sensing-technologies/healthcare-sensing/humidity-with-temperature-sensors/hih-4000-series. Accessed 19 June 2023
21. K. Sasake, View Factor Calculator: Analytical and Monte Carlo Method. https://kanamesasaki.github.io/viewfactor/. Accessed 26 July 2023

Chapter 6
Comparative Assessment of Winter Night Sky Brightness in Southwestern Ontario

William Lubitz, Heather Henry, Alex Nauta, Syeda Tasnim, and Thomas Graham

Abstract Night sky brightness is increasing due to human activity in many parts of the world. There are specific concerns about increasing night sky brightness in Essex County, Ontario, Canada due to increasing use of large-scale supplemental lighting in commercial greenhouses. Increasing concerns about sky brightness have led to implementation of by-laws during the past few years that require greenhouses to fully or partially close light-abatement curtains when supplemental lights are used at night. The sky brightness at a particular location is strongly dependent on all anthropogenic light sources in the area, as well as cloud cover, weather, and moon phase. A longitudinal sky brightness study was completed to investigate the impact of these factors on the sky brightness. Four different locations were identified for long-term sky brightness studies: (1) a "remote" rural location, (2) an urban location in the city of Windsor, (2) a small town location within the town of Leamington, and (4) a rural area with several nearby greenhouses and urbanization. Near-continuous sky brightness measurements were collected between fall 2022 and spring 2023 at each location using Unihedron sky quality meters (SQMs), a widely used sensor for assessing sky brightness. Measured sky brightness is compared between sites. The impact of cloud and moon conditions are assessed: both are shown to contribute to overall sky brightness. Finally, observations are presented on the relative brightness observed between the four sites, particularly in clear or overcast conditions.

W. Lubitz (✉) · A. Nauta · S. Tasnim
School of Engineering, University of Guelph, 50 Stone Rd E, Guelph, ON N1G 2W1, Canada
e-mail: wlubitz@uoguelph.ca

H. Henry · T. Graham
School of Environmental Sciences, University of Guelph, 50 Stone Rd E, Guelph, ON N1G 2W1, Canada

D. S-K. Ting and A. Vasel-Be-Hagh (eds.), *Engineering to Adapt*, Springer Proceedings in Energy, https://doi.org/10.1007/978-3-031-47237-4_6

6.1 Introduction

Night sky brightness is increasing due to human activity in many parts of the world. Anthropogenic light (from buildings, vehicles, roads, etc.) can be scattered back towards the ground by air molecules, clouds, and airborne contaminants, causing the sky to appear brighter. It is well documented that large anthropogenic sources of light (such as greenhouses) can increase localized sky brightness to levels much higher than would exist with only natural light sources [10].

Essex County, in southwestern Ontario, Canada, hosts the largest concentration of commercial greenhouse agriculture in North America, producing large quantities of tomatoes, cucumbers and peppers as well as other crops. The number of greenhouses, and greenhouse production, are both growing in this region, and there are an increasing number of greenhouses located close to the Town of Leamington and several smaller towns. In recent years, more of these greenhouses are moving to year-round production using supplemental lighting to support crop growth during the winter. This has resulted in increasing concerns in the local area about light pollution and night sky brightness due to increasing supplemental lighting in commercial greenhouses. In the past few years, by-laws have been implemented that require greenhouses to partially or fully close light-abatement curtains when supplemental lights are used at night [3, 7]. In response to these bylaws, many (but not all) greenhouses have implemented light abatement curtains and modified their growing practices.

Recent studies by the research team measuring light emissions from Ontario greenhouses on individual nights confirmed that light abatement curtains can reduce light emissions from a greenhouse by two orders of magnitude, and many greenhouses now utilize these curtains [13]. These studies also observed a wide range of sky brightness impacts due to lighting and curtain use at specific greenhouses. Closing curtains sometimes markedly reduced sky brightness at a greenhouse, while at other times, the impact of curtain closure on sky brightness was variable or negligible. This is because sky brightness at a particular location is strongly dependent on all anthropogenic light sources in the area, as well as cloud cover and moon phase and elevation. The prior studies by the research team were only able to measure sky brightness on single nights, which meant it was not possible to examine the effects of things like weather or moon phase [12]. This current study was conducted specifically to get a better view of how different cloud conditions and moon conditions might impact sky brightness in an area.

6.2 Methods

A longitudinal sky brightness study was planned to examine the impact of cloud condition and moon condition on sky brightness in a region. Four different locations (Fig. 6.1) were identified for long-term sky brightness studies:

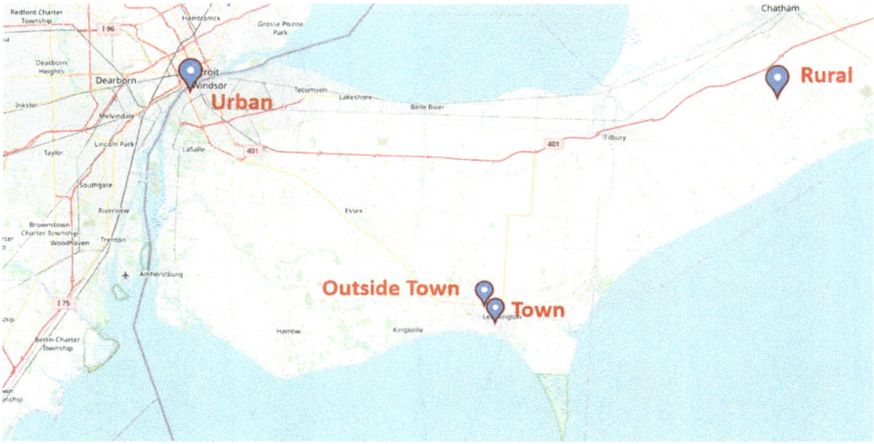

Fig. 6.1 Measurement site locations

1. "Rural": rural location with as little anthropogenic light as possible
2. "Urban": urban location in the Windsor/Detroit urban area, a large population center [14]
3. "Town": within the town of Leamington, a small population centre
4. "Outside Town": a rural area outside Leamington with several nearby greenhouses using supplemental lighting. Note that this site was intentionally not located at or immediately adjacent to any particular greenhouse site.

For this study a sky quality meter (SQM) was installed at each of these four sites between November, 2022 and April, 2023. SQMs (Unihedron, Grimsby, ON, Canada) measure total night sky brightness (including starlight) within a specific field of view. The SQM-UL-DL model used in this study [15] has a 20° field of view and incorporates a datalogger in a weatherproof housing (Fig. 6.2). For sky brightness observations, it is mounted in an area without any immediate sources of light or glare and oriented so that its view is straight upward.

The SQM records sky brightness in units of magnitudes/arcsecond^2, which is a nonlinearly scaled unit. Higher values of sky brightness correspond to darker sky conditions, so a sky brightness of 20 mag/arcsec^2 is a much darker sky than one with a sky brightness of 15 mag/arcsec^2. A decrease in 5 mag/arcsec^2 represents a 100 times increase in observed light intensity. Sky brightness values in mag/arcsec^2 can be converted to the linear units of candelas/m^2 using [6, 15]:

$$\left[\frac{cd}{m^2}\right] = 10.8 \times 10^4 \times 10^{\left(-0.4 \times \left[\frac{mag}{arcsec^2}\right]\right)} \tag{6.1}$$

The sensor in the SQM is a crystalline silicon photodiode TSL237 produced by Texas Advanced Optoelectronic Solutions (TAOS). It is a visible light sensor. The

Fig. 6.2 SQM installed in the field (outside town site)

photodiode is covered by a color-compensating filter HOYA CM-500 with spectral transmittance between 350 and 750 nm. Onboard circuitry converts the current produced by the photodiode to an output frequency proportional to intensity of light incident on the sensor [9].

Unihedron, the manufacturer of the SQM, reports that the SQM has a practical range of 7 mag/arccsec^2 (almost daylight) to 23 mag/arccsec^{2-}(darker than likely to be observed in most locations on earth). SQM manufacturer-reported measurement uncertainty is $\pm$0.1 mag/arccsec^2 [12, 15]. All sensors tested by [9] were found to be within stated linearity uncertainty even at lower light levels, but there were large deviations between individual sensors when measuring the same level of irradiance. Specifically, the 16 sensors (deployed on-site from zero to up to six years) measured a mean of 16.47 mag/arccsec^2 with standard deviation of 0.23 mag/arcsec^2 [9]. There was no consistent pattern of measurement error as a function of deployment time. Comparative testing of the SQM units used in this study showed they were also operating within the manufacturer's stated uncertainty [12].

SQMs are popular for night sky brightness studies [2] in fixed locations and are often used to build night sky brightness maps of a study area to assess spatial light pollution [4, 10]. SQMs have also been attached to various land and water vehicles [1, 5, 8] to obtain spatial night sky brightness data along specific routes.

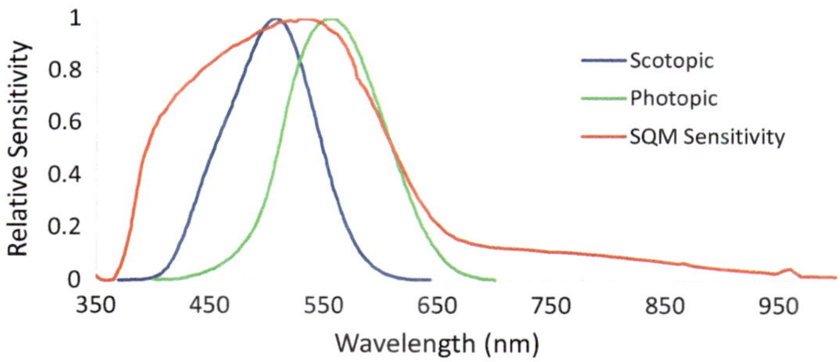

Fig. 6.3 Spectral sensitivity of SQM compared to typical human vision (scotopic and photopic)

Pravettoni et al. [9] measured linearity of 16 Unihedron Sky Quality Meter (SQM) sensors (14 SQM-LE, 2 SQM-LR), over brightness ranging 7.5 orders of magnitude from a mW/cm^2 down to fractions of nW/cm^2 (which corresponds to simulated brightness from 6 to 19 mag/arcsec^2). The testing laboratory was ISO 17025 accredited for electrical testing, qualification and type approval of solar photovoltaic modules, and was modified for the low light levels needed to test the SQM sensors. Because of spectral differences between sun light, and the night sky, the calibrations were relative to other SQM sensors only.

The SQM is a reliable, accurate [9, 11] inexpensive sky brightness meter with a long history of use, but there are several important limitations. The SQM is a single channel sensor intensity sensor, so it does not account for spectral (colour) differences. Several lighting technologies are used in greenhouse lighting, including in Essex County: high pressure sodium (HPS) lights typically produce a yellow or orange light, while increasing popular LEDs can commonly produce a wide range of spectra, including individual colors like red or blue, full white spectrum light, or a range of other options. As a single channel sensor with a fixed spectral response (Fig. 6.3), the SQM cannot differentiate spectrum information, and may record intensity of different colours somewhat differently.

6.3 Data Collection

All four SQMs recorded continuously with a 10 min sampling interval between November 2022 to April 2023. The research team also manually logged information on local weather to be used to identify times of potential SQM contamination by snow or ice accumulation. For analysis purposes, NOAA weather data was sourced from Virtual Crossing (www.visualcrossing.com) for each location. The moon phase and moon elevation were calculated by the SQM software and confirmed using alternative sources.

All four sites were visited several times during the experiment to download data and change batteries. Once data collection was complete, all data was subjected to a quality control process. A data set for analysis was produced by removing data when one or more sites did not record reliable data. Data was also removed if it was possible (based on weather logs) that one or more SQMs could have been contaminated by snow or ice. (For most of the study period there was relatively little snow.)

The dataset was further filtered to only include data representing night conditions, which was taken to be 2 h after local sunset to 2 h before local sunrise. The resulting dataset included 6769 simultaneous observations at the four sites, from Nov. 21, 2022 through April 27, 2023, with notable data gaps in Dec. 2022 (2 weeks) and Feb. 2023 (4 weeks).

6.4 Results

Table 6.1 shows mean observed sky brightness at each site, for the entire data set, and also for four specific time periods during the data collection period. Some of these time periods were associated with significantly more cloud cover than other periods. Table 6.2 shows the mean cloud cover for each site, as reported in the weather data.

Figures 6.4, 6.5, 6.6 show examples of measured sky brightness as a function of time. Each plot shows a four day period, including day time measurements. Note that while the day time measurements (from 2 h before sunrise to 2 h after sunset) are shown in Figs. 6.4, 6.5, 6.6, this data is not included in any of the quantitative analyses.

Table 6.1 Mean observed sky brightness in (mag arcsec^{-2}) for each site, by date range

Mean sky brightness (mag arcsec^{-2})	All data	Nov. 21–Dec. 22	Jan. 2–Feb. 2	Feb. 28–Mar. 28	Mar. 28–Apr. 27
Urban	15.5	15.4	14.9	15.5	16.4
Rural	19.0	19.2	18.6	18.7	19.5
Town	16.0	16.1	14.7	16.1	17.4
Outside Town	15.5	15.5	14.3	15.4	17.2

Table 6.2 Mean reported cloud cover (in % of sky covered by cloud) for each site, by date range

Mean cloud cover (%)	All data	Nov. 21–Dec. 22	Jan. 2–Feb. 2	Feb. 28–Mar. 28	Mar. 28–Apr. 27
Urban	68	73	84	67	42
Rural	52	57	70	47	30
Town	67	74	81	69	39
Outside Town	67	74	81	69	39

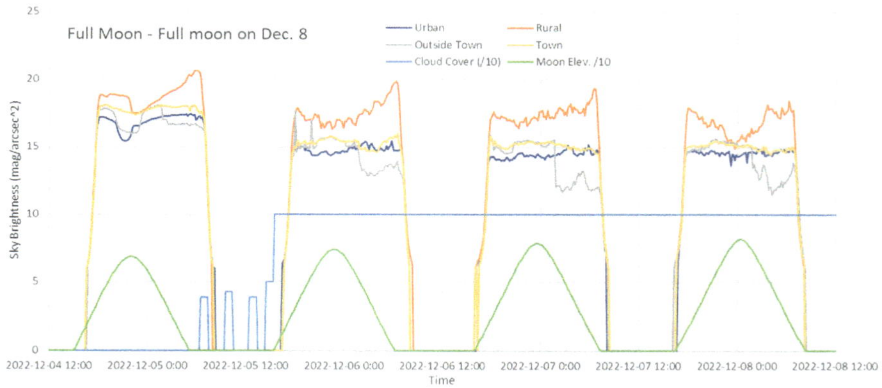

Fig. 6.4 Measured sky brightness Dec. 4–8, 2022 during a bright moon period. Full moon on Dec. 8

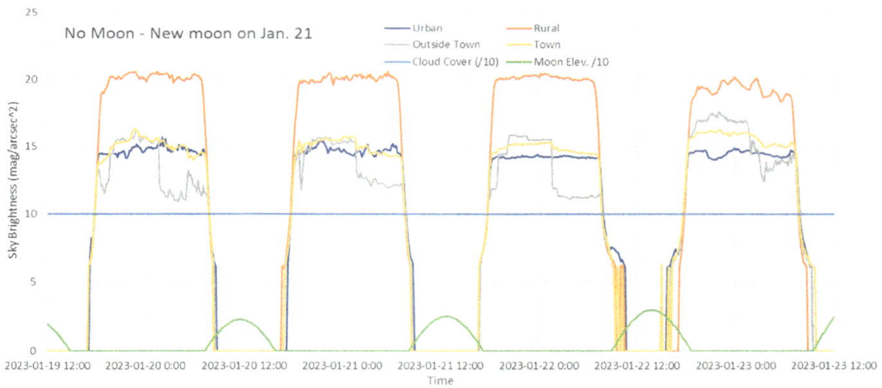

Fig. 6.5 Measured sky brightness Jan. 19–23 during a period with little moon illumination. New moon on Jan. 21

The moon has a notable effect on observed sky brightness. Figure 6.4 shows nights with a full moon, while Figs. 6.5 and 6.6, show nights when the moon is both new and generally below the horizon at night. Comparing figures, there is a clear pattern of increasing sky brightness (low mag/arcsec^2) with increasing moon elevation. It is most visible at the Rural site, where there is little anthropogenic light. Interestingly, there is still brightness from the moon on overcast nights (when cloud cover is 100%, or 10/10 as plotted): in Figs. 6.5 and 6.6, typical Rural sky brightness is on the order of 20 mag/arcsec^2, while on overcast days in Fig. 6.4 values are much less (corresponding to the sky being brighter) and a dependence on moon elevation is visible.

Many Essex County greenhouses turn lights on at 2:00 a.m (although individual greenhouses often use a wide range of lighting timing and protocols). Notably,

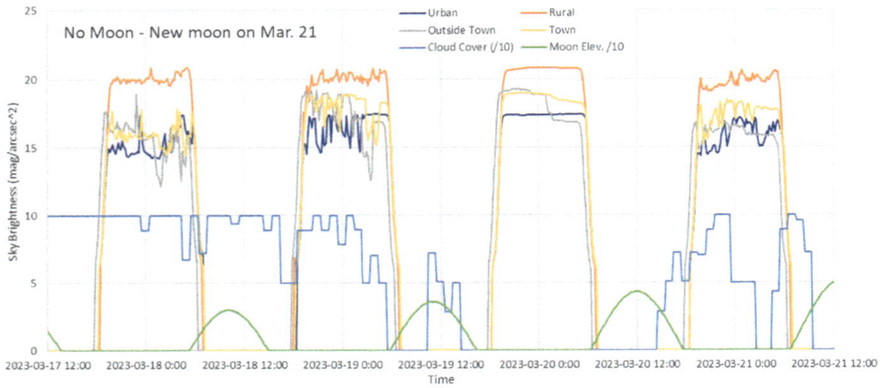

Fig. 6.6 Measured sky brightness Mar. 17–21 during a period with little moon illumination. New moon on Mar. 21

changes in sky brightness at 2:00 a.m are visible in the data on almost all days tested (Figs. 6.4, 6.5, 6.6).

This phenomenon was examined further by examining the extreme cases of entirely clear sky or entirely overcast conditions. There were 345 ten minute records within the data in which all four sites had a reported cloud cover of 0%: these were classified as the "clear sky" case. Another 1595 records included reported cloud cover of 100% at all four sites: there were categorized as the "overcast" case.

For the clear sky case, the average sky brightness at the Town site was less than at the Outside Town site by 0.18 mag/arcsec^2 before 2:00 a.m, however, this average difference increased to 1.07 mag/arcsec^2 after 2:00 a.m. When the sky was overcast, the differences were greater (meaning the sky was brighter at the Outside Town site than at the Town site). Before 2:00 a.m, the Town site was on average 0.20 mag/arcsec^2 than the Outside Town site, while after 2:00 a.m, the average difference increased to 2.38 mag/arcsec^2.

The Outside Town site was on average brighter than the Town site, and the amount of the difference increased after 2:00 a.m (presumably when more greenhouse lighting was turned on). An important follow-on question would be whether the sky brightness above the Town site was increasing in magnitude? This can be examined by using the Rural site—which is far from any artificial lights—as a baseline. Table 6.3 shows average sky brightness difference between the Rural site and each of the others. The same clear and overcast data is used again, and average differences from the Rural site are shown for both before and after 2:00 a.m. (All brightness values are reported as Rural minus the other site, which in practice keeps differences as positive values.)

Table 6.3 affirms that for all cases, the average sky brightness was less at the Rural site than at any of the other sites. Additionally, the sky is much brighter at all three non-Rural sites during overcast conditions, suggesting a significant impact from anthropogenic light reflecting off clouds. If it is assumed that most of the change

Table 6.3 Average difference in sky brightness between rural and other locations (rural minus other) for clear and overcast sky conditions

Sky condition	Case	Outside town	Town	Urban
Clear Sky	After 2:00	2.94	1.87	2.74
	Before 2:00	1.95	1.77	2.97
	Difference	0.99	0.10	-0.23
Overcast	After 2:00	6.28	3.90	4.39
	Before 2:00	3.62	3.42	4.42
	Difference	2.65	0.47	-0.04

in sky brightness at 2:00 a.m is due to greenhouse supplemental lighting coming on at 2:00 a.m in areas outside of Town, some interesting patterns are apparent in Table 6.3. As before, the sky is brighter after 2:00 a.m at both the Outside Town and Town sites, with much larger brightness changes occurring at the Outside Town site. On overcast nights, sky brightness in Town still increases after 2:00 a.m, but only by a fraction as much as it does at the Outside Town site. On clear nights, the average difference in brightness between before and after 2:00 a.m is an almost negligible 0.10 mag/arcsec^2, while Outside Town it remains a noticeable 0.47 mag/arcsec^2.

Table 6.3 also suggests that when anthropogenic ground-based light sources (whether from greenhouses, urban lighting, or other sources) are a dominant source of sky brightness, increasing cloud results in greater amounts of light being reflected back to observers on the ground, significantly increasing observed sky brightness.

Finally, there was little appreciable change in sky brightness at the Urban site after 2:00 a.m during overcast conditions, and only a small change (difference of -0.23 mag/arcsec^2) when the sky was clear, which suggests little impact at the Urban site from the sources that illuminate at 2:00 a.m.

The limitations of this exercise should be noted. While it was observed that multiple greenhouses did begin lighting at 2:00 a.m during several nights, and few other anthropogenic sources were observed or known to start or end at 2:00 a.m, it was not possible to do a definitive survey of this phenomena or quantify the amount and spatial distribution of lighting (greenhouse or other sources) that typically started at 2:00 a.m.

Readers are further cautioned not to over-interpret the data or results. It can be seen (in Figs. 6.4, 6.5, 6.6) that the moon also increases sky brightness when it is present, in both clear and cloudy conditions: this effect would be difficult to quantify and remove. The sky brightness results presented here are a case study: results would likely vary if sensor locations were moved to different locations within the Town, Urban area, or in regions near greenhouses or other light sources. Readers are also cautioned that it would be essentially impossible to identify the contribution of a single facility or light source to regional sky brightness, unless that source was essentially the only light source. The study region contains many greenhouses, other industrial facilities, and commercial and residential developments, all of which produce light emissions that contribute to local sky brightness.

6.5 Conclusion

This study deployed four Sky Quality Meters (SQMs) to examine sky brightness at four different locations from an urban area to a rural area with little anthropogenic light. The results of this study (Tables 6.2 and 6.3, Figs. 6.4, 6.5, 6.6) suggest that supplemental greenhouse lighting is likely affecting sky brightness in the immediate area, and (to a lesser extent) in nearby towns. Effects of greenhouse supplemental lighting were not observed at more distant rural and urban control sites. Cloud cover was seen to increase the sky brightness at all locations with significant anthropogenic light sources (i.e. all sites other than the rural site) due to reflection of anthropogenic light back towards the surface. These conclusions are primarily based on observations of sky brightness variation before and after 2:00 a.m local time, a common time for local greenhouses to initiate supplemental lighting. This is an apparent relationship between some possible factors rather than a definitive demonstration that greenhouse light emissions specifically impacted sky brightness. Future plans include repeating similar comparative measurement with simultaneous ground-truthing and more detailed weather observations, with the goal of being able to separately quantify the impact of anthropogenic and natural light sources on sky brightness in the region.

Acknowledgements The authors wish to thank several site owners for hosting the SQM sensors, without whom this project would not have been possible. We thank William "Danny" Lubitz, Andrew Jenner and Dr. Bill van Heyst at the University of Windsor for supporting the SQM placement at the Urban site. This project was completed as part of a larger research program supported by funding from the Ontario Ministry of Agriculture, Food and Rural Affairs (OMAFRA) Agri-Food Innovation Alliance program.

References

1. X. Ges, S. Bará, M. García-Gil, J. Zamorano, S.J. Ribas, E. Masana, Light pollution offshore: Zenithal sky glow measurements in the Mediterranean coastal waters. J. Quant. Spectrosc. Radiat. Transfer **210**, 91–100 (2018)
2. A. Hänel, T. Posch, S.J. Ribas, M. Aubé, D. Duriscoe, A. Jechow, Z. Kollath, D.E. Lolkema, C. Moore, N. Schmidt, H. Spoelstra, Measuring night sky brightness: methods and challenges. J. Quant. Spectrosc. Radiat. Transf. **205**, 278–290 (2018)
3. Kingsville, Corporation of the Town of Kingsville By-law 96–2020. By-law to Prohibit and Regulate Public Nuisances related to odours and lighting from the cultivation of plants within the Town of Kingsville and to create certain penalties related thereto (2020). Available at: https://www.kingsville.ca/en/town-hall/resources/96-2020-Nuisance-By-law.pdf. Accessed 4 Aug 2023
4. D. Karpińska, M. Kunz, Light pollution in the night sky of Toruń in the summer season. Bull. Geogr. Phys. Geogr. Ser. **17**(1), 91–100 (2019)
5. Y. Katz, N. Levin, Quantifying urban light pollution—a comparison between field measurements and EROS-B imagery. Remote Sens. Environ. **177**, 65–77 (2016)

6. C. Kyba, K.P. Tong, J. Bennie, I. Birriel, J.J. Birriel, A. Cool, A. Danielsen, T.W. Davies, P.N.D. Outer, W. Edwards, R. Ehlert et al., Worldwide variations in artificial skyglow. Sci. Rep. **5**(1), 1–7 (2015)
7. Leamington, Corporation of the Municipality of Leamington By-law 79-20. By-law requiring the abatement of interior greenhouse light (2020). Available at: https://www.leamington.ca/en/municipal-services/resources/79-20-Greenhouse-Light-Abatement.pdf. Accessed 8 Aug 2023
8. M. Liu, W. Li, B. Zhang, Q. Hao, X. Guo, Y. Liu, Research on the influence of weather conditions on urban night light environment. Sustain. Cities Soc. **54**, 101980 (2020)
9. M. Pravettoni, D. Strepparava, N. Cereghetti, S. Klett, M. Andretta, M. Steiger, Indoor calibration of sky quality meters: linearity, spectral responsivity and uncertainty analysis. J. Quant. Spectrosc. Radiat. Transf. **181**, 74–86 (2016)
10. C.S.J. Pun, C.W. So, Night-sky brightness monitoring in Hong Kong. Environ. Monit. Assess. **184**(4), 2537–2557 (2012)
11. S. Schnitt, T. Ruhtz, J. Fischer, F. Hölker, C. Kyba, Temperature stability of the sky quality meter. Sensors **13**(9), 12166–12174 (2013)
12. B. Snow, MASc Thesis. University of Guelph. Measurement and Analysis of Exterior Light Emissions from Commercial Greenhouses (2022)
13. B. Snow, W. Lubitz, S. Tasnim, T. Graham, D. Llewellyn, F. Al-Daoud, Experimental measurements of light emissions from Ontario greenhouses using supplemental lighting at night, in *5th International Conference of the International Commission of Agricultural and Biosystems Engineering (CIGR)*, Quebec, Canada (2021)
14. StatsCan, Population Centre and Rural Area Classification 2016. Statistics Canada (2016). Available at: https://www.statcan.gc.ca/en/subjects/standard/pcrac/2016/introduction. Accessed 4 Aug 2023
15. Unihedron, SQM-LU-DL Operator's Manual (2022). [pdf] Available at: http://www.unihedron.com/projects/darksky/cd/SQM-LU-DL/SQM-LU-DL_Users_manual.pdf. Accessed 18 Aug 2022

Chapter 7
Feasibility Analysis of Solar Tracking Technologies Connected to Renewable Energy Systems

Reza Babaei, David S.-K. Ting, and Rupp Carriveau

Abstract The environmental benefits of photovoltaic (PV) panels have attracted attention worldwide although they still lack the necessary levels of performance and financial viability. To meet these standards, PV tracking technologies are utilized to maximize the efficiency of PV systems. The principal goal of this research work is to perform a techno-economic feasibility analysis and optimal sizing of renewable energy systems integrated into solar tracking technologies for a rural residential user in South Africa. The vertical axis (VC), dual-axis tracker (DA), horizontal axis with monthly (HM), and continuous (HC) adjustments are the tracking options implemented to determine a profitable energy system. The preferred optimal cases based on these scenarios are introduced, and then sensitivity analysis is carried out to estimate the effect of variation of technical, economic, and climatic parameters. The VC-based system offers a financially desirable alternative having net present cost (NPC), levelized energy cost (LCOE), and CO_2 emissions of \$13.7k, \$0.258/kWh, and 2.81.1 kg/year, respectively. Sensitivity analysis indicates that an increase in SOC_{min} would raise NPC and CO_2 emissions. The DA and VC trackers, respectively, show the greatest and lowest increases in NPC due to diesel prices increase. In VC and HM-based cases, PV generation and the renewable fraction are more sensitive to albedo (ground reflectance) variation. Comparing the current optimal results with the previous research in this location demonstrates that NPC and LCOE can be reduced by implementing a proper controlling strategy by \$58k and \$0.005/kWh, respectively, to meet the same electrical load.

Keywords PV tracker · Microgrids · Hybrid energy systems · Photovoltaics · Renewable energy

R. Babaei (✉) · D. S.-K. Ting · R. Carriveau
Turbulence and Energy Laboratory, University of Windsor, 401 Sunset Ave, Windsor, ON, Canada
e-mail: babaeij@uwindsor.ca

D. S-K. Ting and A. Vasel-Be-Hagh (eds.), *Engineering to Adapt*, Springer Proceedings in Energy, https://doi.org/10.1007/978-3-031-47237-4_7

Nomenclature

DA	Dual-axis tracker
DG	Diesel generator
F_1	Fuel curve slope (L/h/ output kW or m^3/h/output kW
F_o	Generator fuel curve intercept co-efficient (L/h/rated kW or m^3/h/rated kW)
f_{pv}	PV derating factor (%)
FT	Fixed tilt
GS	Incident radiation at standard test conditions (1 kW/m^2)
GT	Solar radiation incident in the current hour (kW/m^2)
HKT	Hydrokinetic turbine
HM	Horizontal axis
HV	Vertical axis
LCOE	Levelized cost of energy($/kWh)
L_{max}	Absorption width in maximum power
NPC	Net present cost
O&M	Operation and maintenance cost
P_{DG}	Generator output power (kW)
P_{DG}	Rated generator power (kW)
P_{max}	Maximum available power
PV	Solar photovoltaic
Q	Energy available in the storage at the first timestep (kWh)
Q_1	Available energy remained at the first timestep (kWh)
W_{pv}	Peak power output of PV array (kW)
Y_{DG}	Rated capacity of the generator (kW)
η_{DG}	Generator efficiency (%)

7.1 Introduction

Access to electricity is a critical indicator which impacts the social and financial progress of regions [1]. In this regard, energy scarcity, power instability, and electricity outages have grown worldwide, notably in Africa, where high costs and erratically insufficient electricity have severely affected people's lives [2–4].

Among African countries, South Africa has a large number of rural and energy-deprived populations in which electricity is unavailable to around 30% of rural South Africans. To address this issue, renewable energy can be an environmentally friendly and economical approach to deal with outages and shortages of electricity [5]. Figure 7.1 demonstrates electricity access in rural areas in South Africa and (b) South Africa's electricity production by technology category. The electricity generated by coal as a crucial non-renewable source is predicted to decrease by 150 TWh by 2040. Electricity access has been inconsistent during the past decade in Africa although it started to gradually expand in 2016.

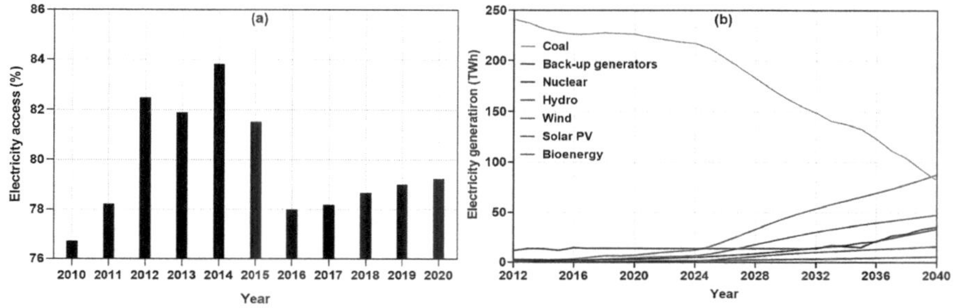

Fig. 7.1 The presentation of **a** electricity access in rural localities and **b** electricity production by technology in South Africa

Solar, wind, geothermal, biomass, and nuclear energy are deemed to be the ideal renewable energy sources for electrification. Another preferable method to provide electricity is to utilize the potential of rivers and oceans, which is known as hydrokinetic (HKT) [6]. With hydrokinetic technology, energy can be converted in a way that has minimal negative effects on the environment. Such as technology has the potential to satisfy the electricity requirements of remote communities, particularly energy-deprived regions in South Africa [7]. The configuration for installing these types of turbines has been demonstrated to be safe and reliable, with minimal environmental impact on the environment. Having mentioned the unique features, there have been a few feasibility research articles on employing micro hydrokinetic power plants to deliver electricity to rural areas.

Despite the described preferred properties of the renewable energy components, the negative of this technology is that the power output varies from hour to hour, resulting in unreliability to supply the load during a day with consistent intensity [8]. Since renewable resources highly depend on meteorological data, and energy storage configuration which makes it challenging to maintain the energy at a certain level of intensity [9, 10].

Some of the previous research articles have investigated the techno-economic feasibility evaluation of these renewable hybrid energy systems (HES) in rural or remote areas. The general category of these studies can be divided into three main groups: optimization algorithm [11–13], design and planning [14–17], comparison procedure [18–20], and energy management [21–23]. For instance, Ref. [24] proposed PV/WT/HKT (Hydrokinetic turbine) system for grid-independent electrification. The findings revealed that this energy solution might be cost-effective and influential in remote regions where grid connectivity is either prohibitively costly or impossible. In Saint Catherine, Egypt, Ref. [25] employed a PV/WT/FC/DG/Hybrid natural gas turbine option to meet a household load of 80 kWh/d. The paper utilized a bi-directional ant colony algorithm to develop a strategy for building optimum systems while minimizing NPC and GHG emissions equally. Ref. [26] also provided an ideal and cost-effective hybrid energy source for a farming community and a suburban neighborhood in a small hamlet in Punjab, Pakistan, utilizing a PV/biomass configuration. The best solution demonstrated energy and net present cost that were both techno-economically realistic, allowing them to be grid-independent. Reference [27] conducted a techno-economic feasibility analysis through PV/DG system for a small-scale load requirement. They stated that a $2,300 salvage cost would result from initial costs of $26,150 being invested. Reference [28] analyzed off-grid PV/WT/DG/biogas generator (BG), and grid-connected PV/WT designs are shown to be the best options. The grid-connected PV/WT was discovered to have a NPC of $6.63M and an energy cost of $0.0234/kWh. The NPC of the hybrid PV/WT/DG/BG is accounted for at 18.1 and $0.1251/kWh, respectively.

The PV system has shown significant improvement and stability in terms of energy generation among the studied renewable energy devices [29–31]. PV systems are becoming more popular owing to their low operating and maintenance costs (O&M) and convenience of usage in addressing rising power demands [8]. PV systems are extremely reliant on local irradiation variations; They are one of the fastest-growing alternatives because they transform solar irradiation into electricity [29]. In this respect, PV tracker techniques can be applied to absorb higher solar energy. Solar tracking system sets the PV collector in the ideal position for receiving the maximum energy [32]. The tracking mechanism can be used as per a single axis (single-axis tracker) or two axes (dual-axis tracker) with dual-axis tracking systems. However, there is an increase in investment and complications in using the two-axis tracking system [33]. In this respect, some previous articles have evaluated the technical, economic, and environmental aspects of tracking technologies integrated into HES. Table 7.1 summarizes the studies that have investigated the optimal sizing and feasibility analysis of PV tracking technology including vertical axis (VC), dual-axis tracker (DA), horizontal axis with monthly (HM), and continuous (HC) adjustments.

When considering the allocation of funds for solar energy systems, the choice between investing in a solar tracker or additional panels emerges as a critical decision with profound implications for overall system performance and financial viability. Several pivotal factors must be taken into account, including solar resource availability, electricity demand profile, and the prevailing costs of solar panels and trackers. Extensive research and studies have substantiated that solar trackers can significantly enhance energy generation, especially in regions with abundant solar irradiance and high energy demand during peak sunlight hours. The dynamic tracking ability of solar trackers optimizes energy capture, leading to a reduction in the number of panels required to achieve the desired energy output, thus not only lowering initial investment costs but also ensuring higher energy yields and better returns on investment [40].

Furthermore, the cost-effectiveness of solar trackers is particularly pronounced in off-grid and remote areas, where optimizing energy production and minimizing expenses are paramount considerations. The continuous adjustment of solar trackers to follow the sun's path ensures increased energy output, even in locations with variable weather conditions. Moreover, ongoing advancements in solar tracking technology and photovoltaic panel efficiencies bolster the case for the adoption of solar trackers as a compelling option. A comprehensive techno-economic analysis, encompassing factors such as the levelized cost of energy (LCOE), government incentives,

Table 7.1 Summary of the previous research papers analyzing impacts of PV tracker devices

Grid connectivity	References	Year	Area	Energy system	Analyzed PV tracker
Off	[34]	2016	Hilly terrain, India	1. Photovoltaic 2. Wind turbine	1. HA 2. Fixed tilt 3. Azimuth tracking 4. DA
	[35]	2017	Eight areas in India	1. Photovoltaic	1. Fixed tilt 2. DA
	[36]	2016	Urumqi, China	1. Photovoltaic 2. Diesel generator	1. Fixed tilt 2. Single axis 3. DA
	[37]	2019	Rural healthcare building, India	1. Photovoltaic	1. HA 2. VA 2. DA
On	[38]	2017	Makkah, Saudi Arabia	1. Photovoltaic	1. HA 2. VA 3. DA
	[39]	2020	Alkharj City, Saudi Arabia	1. Photovoltaic 2. Fuel cell 3. Electrolyzer	1. HA 2. VA 3. DA

and available rebates, empowers system designers to make well-informed decisions aligning with budget constraints, energy objectives, and environmental sustainability goals. By striking an optimal balance between upfront costs and long-term performance, stakeholders can develop economically viable and sustainable solar energy systems, fostering the widespread adoption of renewable energy solutions.

7.1.1 Novelty and Contribution

Examining the previous feasibility assessments of the HESs integrated with PV tracking technologies, the research gaps that this study works on are listed below:

1. There has been no investigation on the compatibility of the HKT-based HES when it is connected to solar tracking systems.
2. The effects of albedo and SOC_{min} on operational performance, financial parameters (NPC and LCOE), and emissions rates were not studied yet.
3. Limited consideration has been drawn into project lifetime effects and technology expense of solar trackers on the cost-effectiveness of the renewable HESs.
4. Very few feasibility evaluations of renewables have taken into account the fluctuation of diesel prices and inflation rates as key financial aspects of the system's energy costs.

Hence, the primary objectives of this study encompass (i) the identification of stable and cost-effective PV tracking technologies, considering meteorological, economic, and renewable data; (ii) a comprehensive comparison of power production, reliability, emissions rates, present and energy costs; (iii) conducting a sensitivity analysis to assess the impact of fluctuating techno-economic inputs on present and energy costs; and (iv) a comparative analysis with the findings of previous research conducted in this area. Furthermore, this study contributes to the field by providing valuable insights and contributions in the following aspects:

1. Introduction of the detailed framework and comprehensive decision-making strategy for optimal sizing and feasibility analysis of HES.
2. This is the first study that investigates the feasibility of solar tracking technologies connected to HKT-based HESs in South Africa.
3. The effects of technical parameters such as albedo (ground reflectance), $SOC_{min,}$ of battery, as well as economic parameters such as capital cost of solar trackers, on the NPC, LCOE will be investigated.

7.2 Methods and Materials

The research adopts a decision-making approach, serving as a comprehensive and extensive framework to guide system designers in achieving economically and technically suitable hybrid energy system (HES) designs and assessing system performance. Specifically, the proposed HES configuration integrates photovoltaic (PV), diesel generator (DG), and battery components to supply power to a rural building in South Africa. The subsequent sections present detailed information pertaining to the study's geographical locations, local resources, employed methodology, and the various energy components involved in the proposed HES.

7.2.1 Demonstration of the Framework Implemented for the Optimal Design of This Study

A thorough review of prior analyses and literature on hybrid energy systems (HES) conducted in diverse global locations has revealed a lack of a comprehensive strategic framework for HES design and planning. Addressing this gap, the present research endeavors to provide a comprehensive and systematic framework for HES design and optimization. The proposed HES framework serves as a decision-making tool to enhance the effective utilization and adaptation of HES solutions. Figure 7.2 illustrates the five key stages encompassing the paper's HES framework. In the initial stage, the primary objectives of HES investments are presented, encompassing the justification for deploying HES in commercial, industrial, and residential sectors. The framework succinctly outlines the main purposes for adopting HES, including:

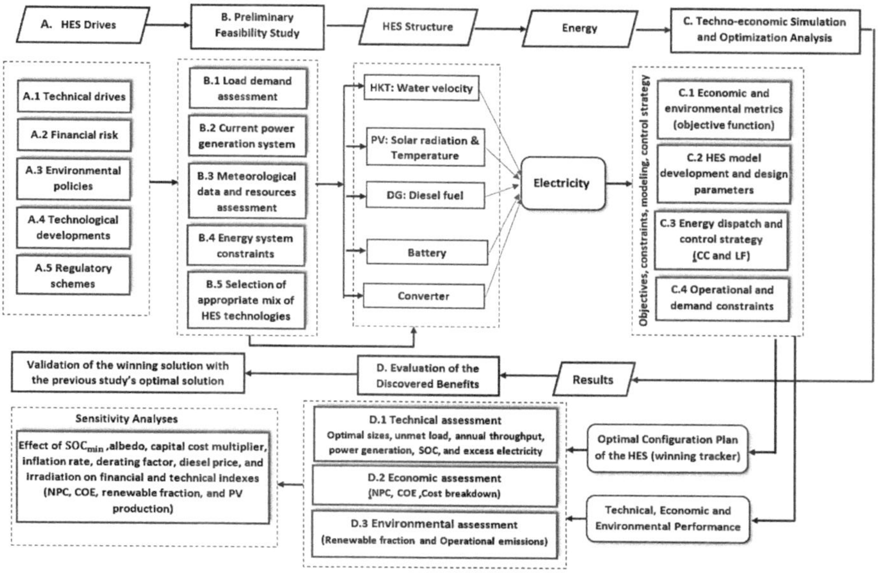

Fig. 7.2 The comprehensive procedure of optimal sizing for hybrid energy systems

1. **Technical motivation**: a suitable HES setup determines multiple technical advantages for utilities and HES users which improve power reliability and provide high-quality power.
2. **Economic challenges**: HES setup is normally installed to satisfy the required load; therefore, it offers advantages to power industry owners by reducing the costs of establishing typical centralized systems, transmission, and distribution lines, and associated infrastructure.
3. **Environmental roadmap**: maximizing the foundation and infrastructure of HES to address climate change and lowering the gas emissions toward reaching zero-emission buildings and sustainable communities.
4. **Technological advancements of components**: HES is regarded as an economical option in the current energy industry, given recent advances and huge progress in renewables and energy storage.
5. **Regulatory plans**: HES offers systems to improve the variety of power sources and boost power security.

The second stage involves conducting a preliminary feasibility analysis, which serves as a pivotal step in selecting appropriate HES technologies and determining an optimal combination to meet the load profile requirements. This stage adopts a foresight approach, necessitating a thorough evaluation of the key input parameters for HES. The primary input parameters under consideration include:

1. The activity of renewable energy resources and meteorological data.
2. Expected load profiles, main categories (residential, commercial, agricultural, etc.), and appliance consumption primarily depend on the occupant type.
3. The present power production system meets load profiles.
4. Energy system constraints related to the HES components can restrict or prohibit the use of some energy options, such as geographic constraints (e.g., merely some area is applicable for the installation of solar panels or challenges limiting the durability of wind turbine foundations) [41]

Once the energy generation options have been identified, the next step involves establishing the most effective controlling strategy. This decision is crucial in optimizing the system and addressing existing challenges. Given the paramount importance of financial considerations in electrification plans, the optimization framework incorporates various HES economic metrics to determine the most suitable solution. Depending on the specific context, such as energy components, geographical data, and financial and technical characteristics, cycle charging (CC) and load following (LF) dispatch strategies emerge as promising methods to significantly reduce cost indexes. Moving to the third step, a comprehensive techno-economic design and optimization study (stage C) becomes essential. This involves accurately defining the technical and economic parameters of the energy solutions to avoid incorrect component sizing. To ensure reliability, an appropriate operational and optimization plan must be maintained, supported by a reliable economic investigator. Consequently, the proposed optimized HES is evaluated simultaneously for its operational, economic, and environmental characteristics. Step D specifies which parameters under technical, economic, and environmental viewpoints will be evaluated. Additionally, conducting sensitivity analysis is crucial to generalize results under varying technological, economic, and climatic conditions. The following sections delve into the implementation of techno-economic optimization and sensitivity evaluations.

7.3 The Intended Site Description

Figure 7.3 depicts a rural family in a distant region of Kwazulu Natal province, South Africa. This place enjoys a humid environment with a geographical elevation of 1,169 m above sea level. The annual building load profile is depicted in Fig. 7.4, which was acquired from Ref. [42], residential power tends to be procured for equipment such as fridges, TV, lights, kettle, toaster, iron, cell phone chargers, etc., with an annual load of 38 kWh/day.

Fig. 7.3 The map of the rural place located in South Africa

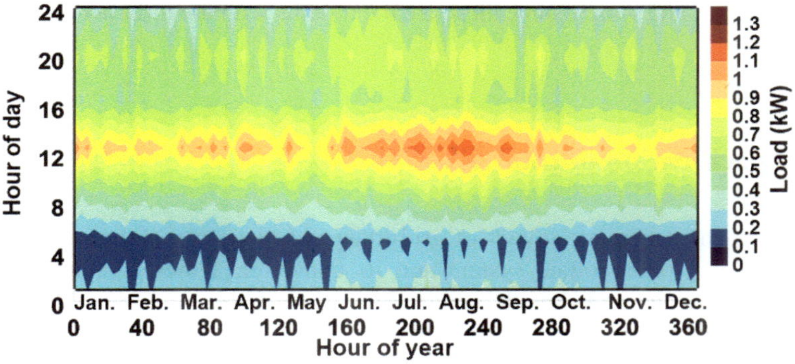

Fig. 7.4 The annual electrical load of the residential place

7.3.1 Renewable Resources

Figure 7.5 depicts the water velocity and solar irradiation from the sun at the intended place. NASA [43] and an experimental investigation in this vicinity [44] provided the monthly data. In comparison to most other places in the globe, irradiation from the sun in the summer fluctuates between 3.5 to 5.5 kWh/m^2/day. The highest and lowest irradiation is observed in December and June, respectively. The decrease in solar irradiance in the summer period can be attributed to factors such as seasonal variation in sun angle, weather patterns and cloud cover, regional climate, and the impact on solar energy systems in this area. The average monthly water velocity profile confirms that the hydrokinetic turbine would produce the highest energy

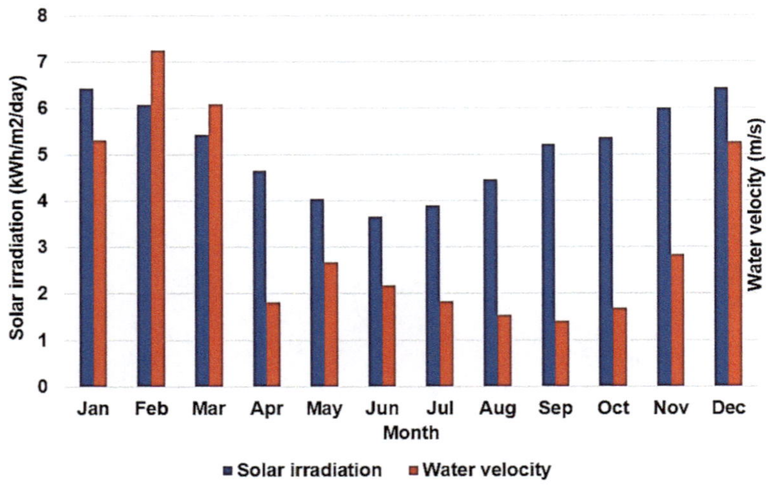

Fig. 7.5 Representation of monthly **a** water speed and **b** solar irradiation [51]

yield in the fall and winter since these are the months with the most accessible water velocities.

7.4 Specification of the HES

In Fig. 7.6, the schematic layout of grid-independent energy systems is depicted, consisting of a photovoltaic system and its converter, a diesel generator, and a hydrokinetic turbine, all directly connected to the AC bus. Additionally, batteries are integrated into the system, linked to the AC through an AC/DC bidirectional converter to facilitate power-sharing. Table 7.2 presents the financial attributes and expected lifetime of each energy production component.

7.4.1 Diesel Generator

Diesel generators in off-grid solutions satisfy the electrical load first, which reduces the reliance on batteries, resulting in cost—savings. The generator efficiency is shown in the formula below [50]:

$$\eta_{DG} = \frac{3.6P_{DG}}{GHC.LHV_{D}} \tag{7.1}$$

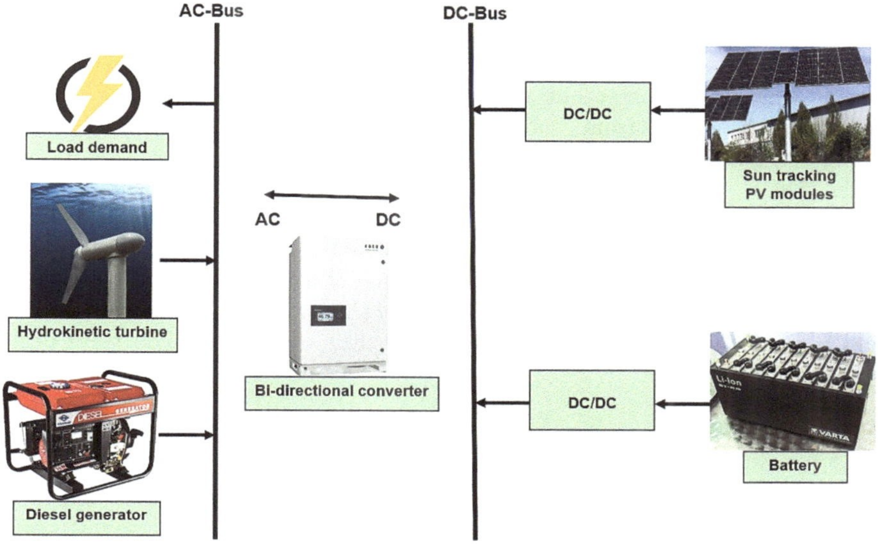

Fig. 7.6 The intended hybrid energy solution of the schematic configuration

Table 7.2 Financial description of the energy components

Cost type	Energy component				
	PV system	Diesel generator	Hydrokinetic turbine	Battery	Converter
Capital expense ($/kW)	550	350	1,300	200/kWh	300
Replacement expense ($/kW)	550	325	870	180/kWh	300
Lifetime (yr)	25	20	15	15	15
Reference	[45]	[46]	[47]	[48]	[49]

where η_{DG}, ρ_f, and P_{DG} stands for the generator efficiency (%), the output power (kW), the fuel density (kg/m^3), and the rated generator power (kW), respectively. The hourly generator's fuel consumption (L/h) is shown in the following equation [49]:

$$\dot{m}_{fuel} = F_0 Y_{DG} + F_1 P_{DG} \tag{7.2}$$

Y_{DG} is the generator's rated capacity in this equation (kW), F_o is the intercept co-efficient of the generator fuel curve (L/h/rated kW or m^3/h/rated kW), and F_1 is the fuel curve slope (L/h/output kW or m^3/h/output kW.

7.4.2 PV System

The effectiveness of solar panels varies depending on the surroundings (temperature in the environment, moisture, and so on), solar irradiation ($kWh/m^2/day$), and technology for collectors. To compute the PV module's output power, use the formula given [51]

Where W_{pv} is the PV array's peak power output (kW), f_{pv} is the PV derating factor (%), the current hour's solar irradiance is denoted by G_T (kW/m^2), The incident radiation at standard test settings is denoted by G_S (1 kW/m^2), α_p is the temperature coefficient (%/°C). T_C stands for the PV module temperature in degrees Celsius (°C) and T_S stands for the PV module temperature in the test condition in degrees Celsius (°C).

7.4.2.1 Solar PV Tracking System Technologies

PV tracking mechanism is utilized to adjust PV modules toward the sun's rays to enhance the PV modules' efficiency and productivity by maximum utilization of solar irradiance. Fixed-tilt solar PV modules typically employ manually adjustable panel slops for simplicity and cost-effectiveness. The PV system is situated at a fixed slope, and azimuth suffers from a considerable deficiency in solar irradiation as a result of the sun passing through the day and its orbit fluctuating annually [29]. Sun-tracking PV systems typically can be constantly adaptable toward high-density rays from the sun. The followings are the main PV tracking methods and corresponding scenarios applied in this analysis:

1. **Scenario I**: Horizontal-axis monthly adjustment (HM). Each month, the tilt angle is changed to ensure that the panels at noon are at almost a perpendicular angle to the sun's beams while the rotating axis circulates around the horizontal (east–west) axis.
2. **Scenario II**: Horizontal-axis continuous adjustment (HC). The tilt angle is unceasingly altered while the rotation is around the horizontal.
3. **Scenario III**: Vertical-Axis continuous adjustment (VC). The tilt is fixed, but the PV panel rotates infinitely around the vertical (north–south) axis.
4. **Scenario IV**: Dual-axis-tracker (DT). To keep a perpendicular angle between PV arrays and sun beams, the PV panels continuously spin in both directions (horizontal and vertical).

These PV tracking technologies are classified based on how many rotational axes they have, as shown in Fig. 7.7.

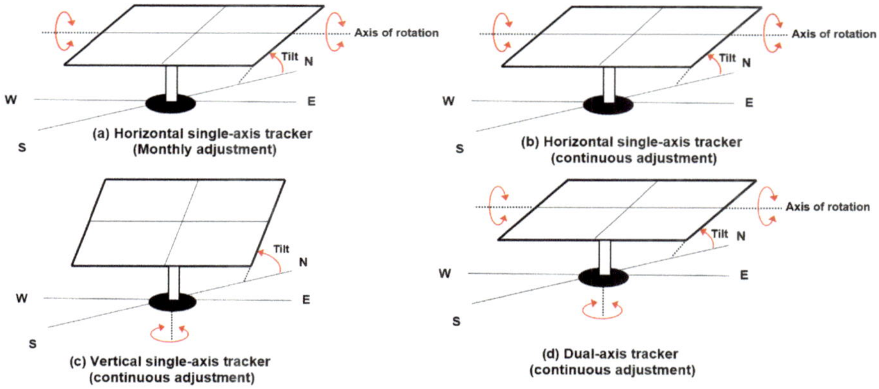

Fig. 7.7 Illustration of investigated tracking systems

7.4.3 Hydrokinetic Turbines

The water's velocity affects the features of hydrokinetic turbines. The longitudinal system absorbs the energy from the waves to create energy (which is placed vertically in the direction of the waves in the water). 3 Hydraulic systems positioned inside the device's joints turn the energy into electricity. Then, using proprietary units and cables, the energy generated is converted to shore power. The largest amount of energy available from sea waves can be seen in the equation below [52].

$$P_{max} = \frac{\rho g^2 T H^2}{64\pi} \times L_{max} \tag{7.4}$$

The maximum available power is P_{max} in this case, ρ is the water density, g is the gravitational acceleration, T is the wave period, The important wave height is indicated by the symbol H, and L_{max} is the maximum power absorption width.

7.4.4 Battery Storage

Li-ion batteries are regarded as a backup because of their dependability, affordable efficiency, high depth of discharge (DOD), and adaptable technology. It aids HES in distributing and preserving energy generated by components so that it can be used in times of uncertainty. Equation (7.5) [53] can be used to estimate the amount of stored energy.

$$P_{Batt,Cmax} = \frac{\min(P_{Batt,Cmax,kbm} P_{Batt,Cmax,mcr} P_{Batt,Cmax,mcc}}{\eta_{Batt,C}} \tag{7.5}$$

In this case, $\eta_{Batt,C}$ is the efficiency of charge storage,

$$P_{Batt,Cmax,kbm} = \frac{kQ_1 e^{-k\Delta t} + Qkc(1 - e^{-k\Delta t})}{1 - e^{-k\Delta t} + c(k\Delta t - 1 + e^{-k\Delta t})} \tag{7.6}$$

$$P_{Batt,Cmax,mcr} = \frac{(1 - e^{-\alpha_c \Delta t})(Q_{max} - Q)}{\Delta t} \tag{7.7}$$

$$P_{Batt,Cmax,mcc} = \frac{N_{Batt}.I_{max}.V_{nom}}{1000} \tag{7.8}$$

The constant of storage rate, $k(h^{-1})$, is used here, at the first timestep, Q_1 represents the amount of energy that was still available. (kWh), Δt is time step duration (h), The amount of energy stored in the batteries is denoted by Q at the initial stage (kWh), c is the ratio of storage capacity, α_c is the maximum charge of the battery (A/Ah), Q_{max} is the total capacity of storage (kWh), N_{Batt} is the quantity of batteries, I_{max} is the battery's maximum battery current (A), and V_{nom} is the battery's nominal voltage.

7.4.5 Converter

To harmonize hybrid energy systems, the system converter is crucial. The main function of a converter is to keep energy flowing between AC and DC. It acts as a medium for converting DC to AC electric power and connects the two systems converting DC to AC. The converters' power rating is derived using Eq. (7.9) [51]:

$$P_{inv}(t) = (\frac{\rho}{\rho_0}).P_{max.L}(t)/\eta_{inv} \tag{7.9}$$

where $P_{max.L}$ is the maximum load demand, and inverter efficiency is defined as η_{inv}.

7.4.6 Financial Equations

The total cost, consisting of initial, replacement, operation, and maintenance (O&M), and fuel costs, minus the project's salvage cost at the end of its life, is expressed as the net present cost (NPC). Equation (7.10) can be used to get the total NPC [14]

$$C_{npc,tot} = \frac{C_{ann,tot}}{CRF(i, R_{proj})} \tag{7.10}$$

where $C_{ann,tot}$ represents the total annual cost (\$/year), the yearly real interest rate is symbolized by the letter i (%), The TP signifies the project's lifetime (year), and

CRF stands for element of capital recovery, This is measured utilizing the equation below [40]:

$$\mathrm{CRF(i, n)} = \frac{\mathrm{i(1 + i)^n}}{\mathrm{(1 + i)^n - 1}} \tag{7.11}$$

The project's lifetime (year) is represented by n, and i is the annual interest rate, which is computed using the equation [19]:

$$\mathrm{i} = \frac{\mathrm{i - f}}{\mathrm{1 + f}} \tag{7.12}$$

where the rate of yearly inflation is (%).

Levelized cost of energy (LCOE): among the most significant measures for evaluating HES's cost-effectiveness. The LCOE is calculated by dividing the average HES cost by the total electrical energy served (kWh), which is measured by Eq. (7.13) [54]:

$$\mathrm{LCOE} = \frac{\mathrm{C_{ann,tot}}}{\mathrm{L_{ann,load}}} \tag{7.13}$$

Here, $\mathrm{L_{ann,load}}$ is the overall yearly electrical consumption (kWh/year), and $\mathrm{C_{ann,tot}}$ is the total annual expense ($/year).

7.5 Results and Discussion

The technical, environmental, and financial results of the winning energy systems are explained in this part and then will be complemented with the sensitivity assessment and a more detailed examination of each target area.

7.5.1 The Outcomes of the System Optimization

HOMER simulates hybrid energy alternatives, and solutions that are not practical (those that fail to meet the user-specified restrictions) will be excluded from the final optimization outcomes and not exhibited. The NPC and LCOE determine how the feasible options are ordered. Table 7.3 presents the technical and financial results of HESs and the considered tracking technologies. The results show that using a VC tracker integrated into DG/PV/HKT results in the least NPC of $13.70 k. LCOE of this solution also is $0.058/kWh less than PV/HKT hybrid system. Implementing a PV/HKT system with a VC tracker reveals the most appropriate option in electricity production. The average of surplus electricity and the capacity shortage is observed to be higher amongst PV/HKT options. The breakdown of the associated expenses in tracker-based optimal scenarios is shown in Fig. 7.8. The majority of the total expenses are dedicated to the initial cost of the components. Moreover, the optimal HV-based HES achieves a $1.52k, $0.55k, and $0.41k higher capital cost than optimal HESs under VC, HM, and DA, respectively. Under the DA tracker, fuel usage and replacement costs are found to be higher than in other cases.

Using DG to produce electricity is damaging to the environment since the burning of diesel produces gaseous pollutants. Gas emissions, including CO_2, NO_x, PM, UHC, SO_2, and CO, are among the pollutants released from diesel generator [55]. Table 7.4 compares the total GHG emissions emitted by the winning energy source using different solar tracking systems from an environmental standpoint. The VC-based option is the most environmentally friendly alternative as compared with other options, emitting the lowest CO_2 level (282 kg/year). Emissions rates from HESs connected to HM and HC trackers are roughly similar.

7.5.2 Techno-Economic Analysis of the Energy Solutions

It is crucial to examine the major financial metrics and power production profiles for optimal HES under each PV tracker. Table 7.5 illustrates the electrical outputs of each component under the integration of intended PV trackers. As per the table, the energy costs of the DG in all tracker options are measured at $0.273/kWh, which is comparable to HKT (0.0213 $/kWh). The greatest energy cost from the PV is seen under the HM tracker system at $0.075/kWh. The least energy costs of PV generation compared with HKT and DG indicate the desirably economical implementation of PV systems in South Africa.

The term “unmet electric load” refers to the portion of the load that remains unsatisfied due to insufficient energy output. Depending on the magnitude of the unmet electric load and renewable fraction, adjustments can be made to the PV size, either increasing or reducing it accordingly. Moreover, the concept of storage throughput (kWh/year) is introduced, representing the total energy flow passing through the battery units annually. The average energy between the incoming and outgoing energy is utilized to calculate the battery throughput, which provides insights into the battery’s operational lifespan. Notably, there exists an indirect relationship between the annual throughput and the battery’s lifetime. As depicted in Fig. 7.9, all the considered options exhibit unmet loads, implying varying levels of reliability in meeting the required load. Among these options, the VC-based DG/PV/HKT configuration and the HC-based hybrid PV/HKT option demonstrate higher reliability due to their lowest unmet loads. Conversely, configurations involving DA trackers exhibit lower reliability in fulfilling the load for both scenarios.

Figures 7.10 and 7.11 illustrate the annual profile of the battery state of charge (SOC) and its relative frequency for the selected winning tracking technology (VC). Throughout the year, with the exception of December, the lowest SOC values are predominantly observed from the beginning of fall to the end of April. This decline

Table 7.3 Results of the optimal hybrid energy systems under the integration of different tracking technologies

Components	Tracker	Controller	NPC	LCOE	PV	DG	HKT	CNV	BT	Electricity produced	Excess electricity
DG/PV/HKT	VC	LF	13.70	0.258	3.05	2	1	6	2.12	7.36	1.98
	HM	LF	15.04	0.283	2.5	2	1	6	2	8.65	2.27
	HC	LF	15.6	0.293	2.3	2	1	6	2.01	9.52	3.15
	DA	LF	16.21	0.286	1.8	2	1	6	2.18	10.4	4.02
PV/HKT	VC	CC	16.8	0.316	5.65	-	1	13	2.9	9.3	5.1
	HM	CC	19.45	0.366	4.5	-	1	16	2.86	10.11	5.44
	HC	CC	20.9	0.393	4.6	-	1	6	2.71	10.88	5.8
	DA	CC	21.18	0.399	4.2	-	1	16	3.48	11.29	6.91

* PV = Photovoltaic system, DG = diesel generator, HKT = hydrokinetic turbine, CNV = converter, BT = battery

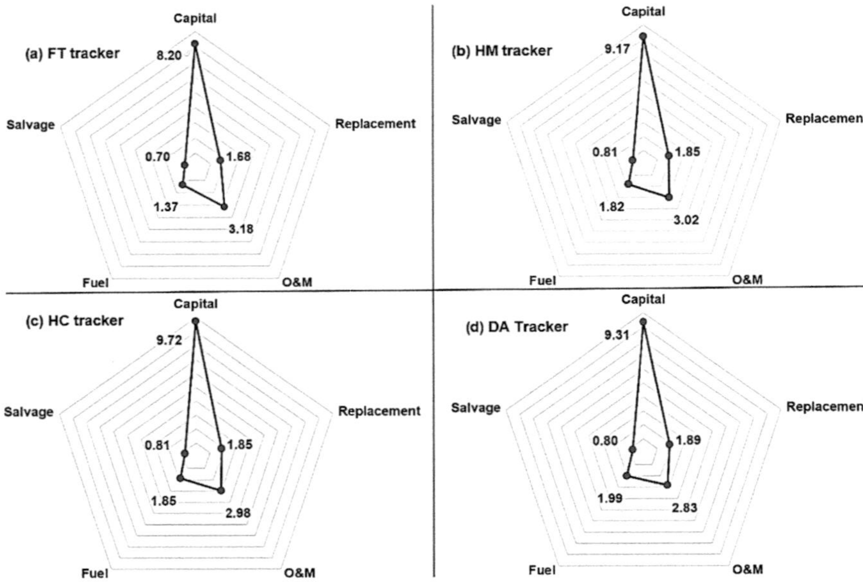

Fig. 7.8 Cost breakdown of each tracker technology connected to the HES

Table 7.4 Breakdown of the emission rates from the tacker-based scenarios

Gas type	Tracking configuration			
	FC	HM	HC	DA
CO_2 (carbon dioxide)	277	369	375	404
CO (carbon monoxide)	1.7	2.3	2.3	2.5
UHC (unburned hydrocarbon)	0.076	0.10	0.10	0.11
PM (particulate matter)	0.010	0.013	0.014	0.015
SO_2 (sulphur dioxide)	0.6	0.9	0.9	0.9
NO_X (nitrogen oxide)	1.6	2.1	2.2	2.3
Total yearly emissions (kg/year)	282	374.4	380.5	410

in SOC can be attributed to the limited and unstable solar irradiation during this period. On the other hand, the SOC levels tend to fluctuate around 58.4% and 100% for approximately 68% of the time each year, indicating that the batteries are mostly operating in a "shallow" discharge state during the majority of the year. This finding suggests that the batteries are adequately charged and capable of meeting the energy demands effectively for a significant portion of the year.

The occurrence of excess electricity results from a discrepancy between energy production and the duration of consumption. As depicted in Figs. 7.12 and 7.13, the surplus energy is mainly generated between 10:00 a.m. and 5:00 p.m., with the peak of extra power observed around noon due to the higher PV output during that period.

Table 7.5 Presentation of the component outputs in considered tracking technologies

	PV production	HKT production	DG production
VC	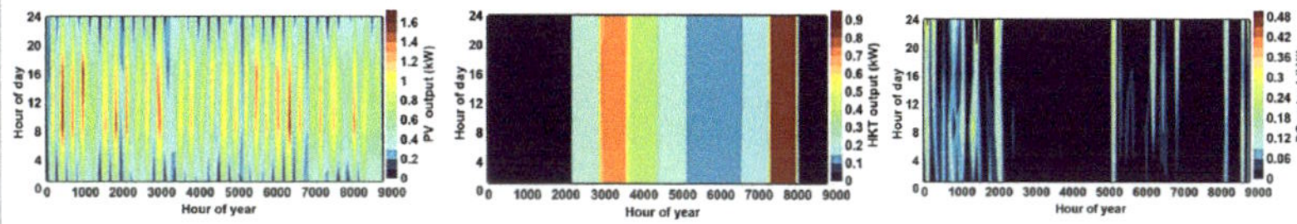		
	Yearly production:5,030 kWh/year Equipment energy cost: $0.042/kWh	Yearly production:2,075 kWh/year Equipment energy cost: $0.215/kWh	Yearly production: 270 kWh/year Equipment energy cost: $0.275/kWh
HM	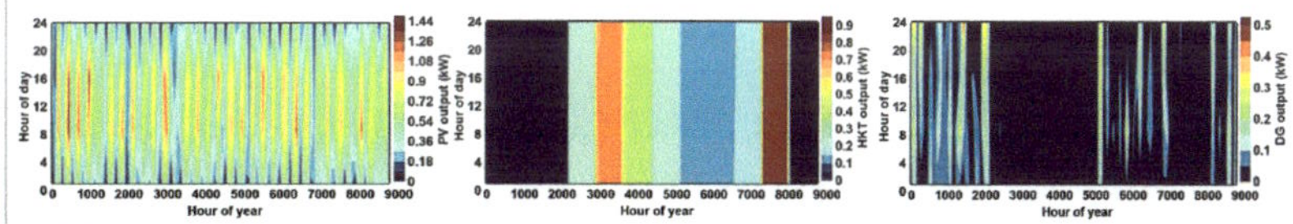		
	Yearly production:4,235 kWh/year Equipment energy cost: $0.063/kWh	Yearly production: 2,085 kWh/year Equipment energy cost: $0.213/kWh	Yearly production: 360 kWh/year Equipment energy cost: $0.273/kWh

(continued)

Table 7.5 (continued)

	PV production	HKT production	DG production
HC	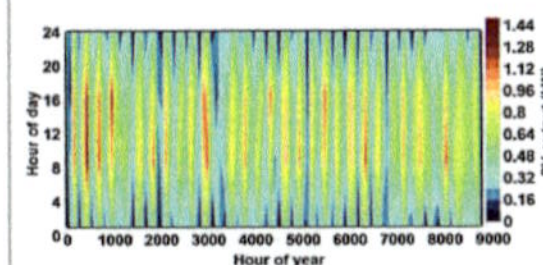	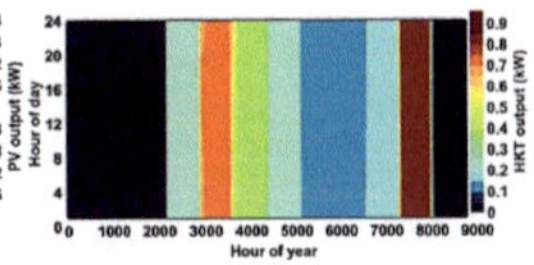	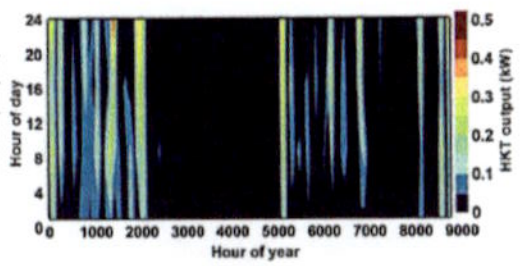
	Yearly production: 4,102 kWh/year Equipment energy cost: \$0.075/kWh	Yearly production:2,082 kWh/year Equipment energy cost: \$0.213/kWh	Yearly production:372 kWh/year Equipment energy cost: \$0.273/kWh
DA	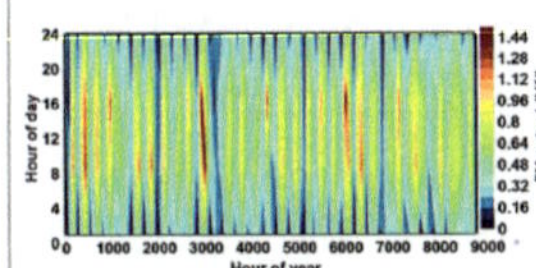	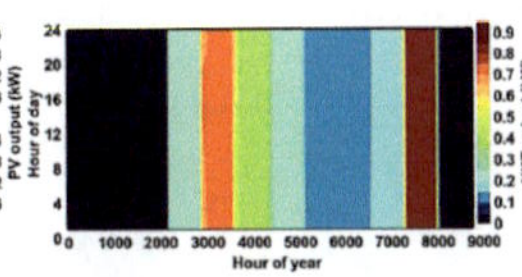	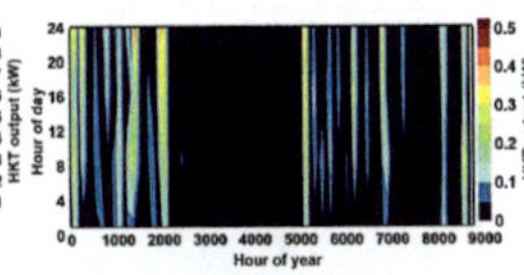
	Yearly production: 3,895 kWh/year Equipment energy cost: \$0.068/kWh	Yearly production: 2,073kWh/year Equipment energy cost: \$0.215/kWh	Yearly production: 402 kWh/year Equipment energy cost: \$0.275/kWh

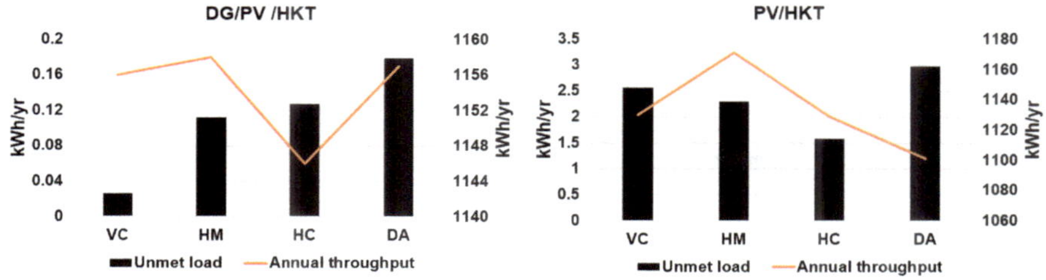

Fig. 7.9 Comparison of unmet load and annual throughput

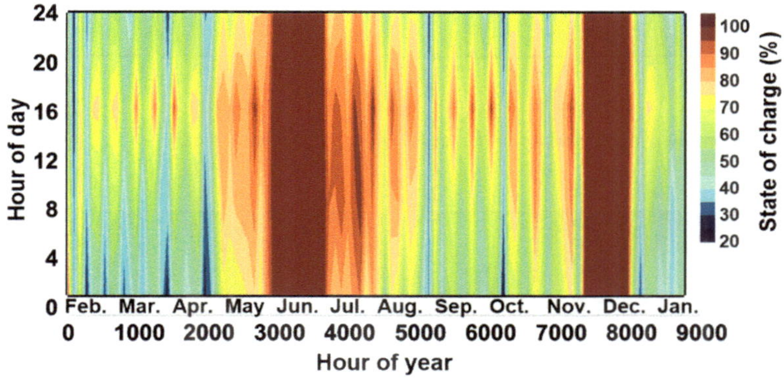

Fig. 7.10 Yearly profile of the battery state of charge (SOC) under the winning system

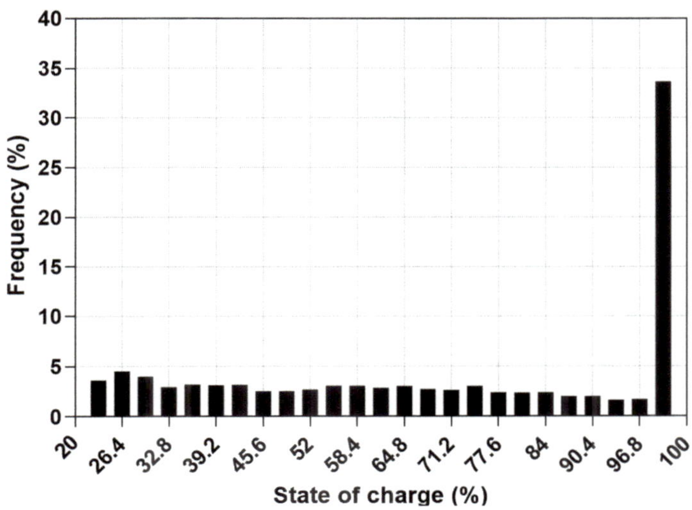

Fig. 7.11 Relative frequency for the SOC under the winning option

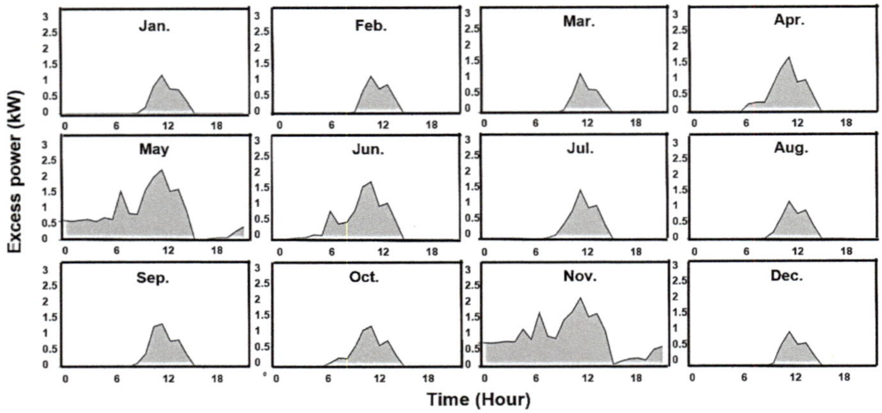

Fig. 7.12 Monthly average variation of excess power in the winning case

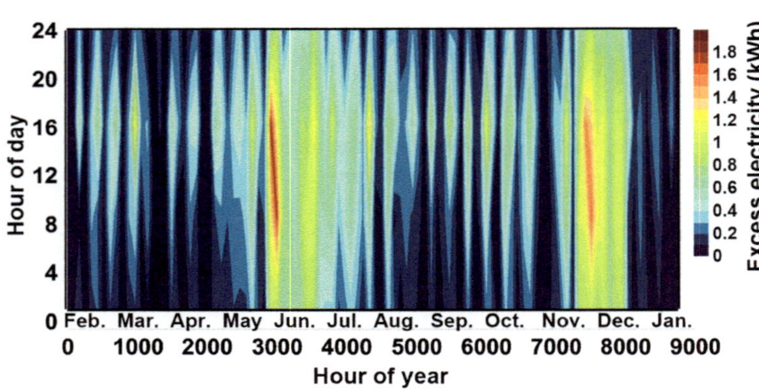

Fig. 7.13 Hourly profile of excess power in the winning case

While the selected HES with VC tracker effectively meets the energy requirements of the site, it is worth noting that during the months of February to May and August to November, it is not the most opportune time to fully charge the batteries. This is likely due to the relatively lower availability of excess electricity during these time periods, which affects the ability to maximize battery charging.

7.5.3 Sensitivity Evaluation

The sensitivity assessment of the winning case is shown in this part. This analysis allows predicting how the energy option would change under various technical and economic circumstances. SOC_{min} is the lowest magnitude of battery charging status

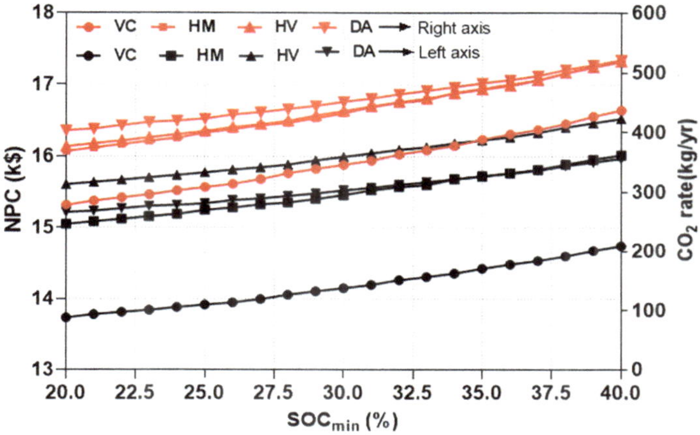

Fig. 7.14 Impacts of SOC_{min} on NPC and CO_2 in various scenarios

below which the battery is never discharged. To prevent the battery bank from being damaged by excessive discharge, the SOC_{min} has not been tuned to a very low amount [56]. Figure 7.14 shows the impacts of the SOC_{min} fluctuation on NPC and CO_2 production. An increase in SOC_{min} leads to growth in NPC and CO_2 production of all tracking technologies. The most significant impact is seen under the VC tracker; a 20% increase in SOC_{min} value will increase NPC and CO_2 by $0.17k and 1.1 t/yr, respectively. As the VC tracker is not able to receive intense irradiation compared to other tracking devices, it will be more reliant on the diesel generator. An increment in SOC_{min} would increase the DG's reliance on it to meet the required load, and then higher O&M expenses, which leads to increasing NPC and CO_2 levels.

Albedo refers to the capacity of an object or surface to reflect light from the sun. As per the meteorological conditions in the case area, the ground cover is potentially snow or rain as well as grass in winter and summer, respectively. As a consequence, the impacts of albedo fluctuation on PV performance and its corresponding expenses are explored in this study. The capacity of an object to reflect light from the sun is known as "albedo." The change of ground reflectance against PV generation and the renewable fraction is depicted in Fig. 7.15. Increased albedo enhances solar irradiation on PV arrays, which results in PV output and renewable fraction. The VC-based and HM-based energy systems show the greatest increase in PV production and RF values. Figure 7.16 represents the variation of albedo on the NPC and LCOE for each hybrid solution. The albedo increase would decrease the required PV size leading to lower NPC and LCOE. The highest reduction in NPC and LCOE is observed under DA-based cases, Fig. 7.17 illustrates how the energy cost of the unit varies when the capital cost of tracking equipment changes. The expense of the DA tracker should decrease by a minimum of 35% in order for the resultant to be comparable to that of HV with the initial value.

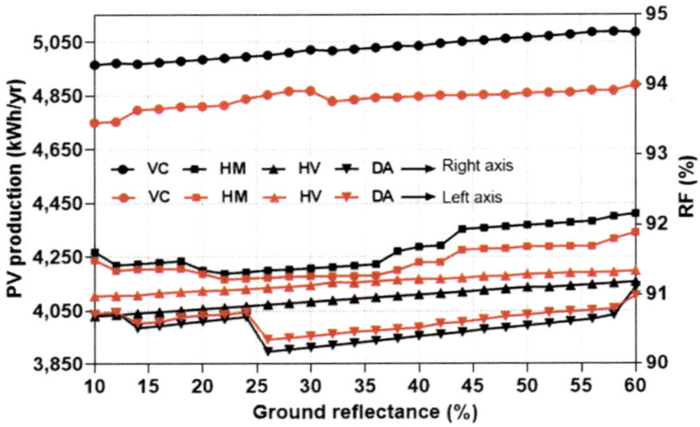

Fig. 7.15 The impact of albedo on PV production and renewable penetration of the considered tracking technologies

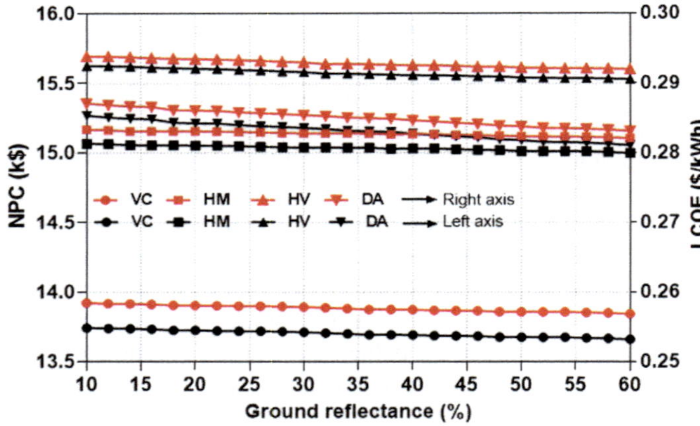

Fig. 7.16 The impact of albedo on NPC and LCOE of the considered tracking technologies

Derating factor of PV panels is another studied sensitivity variable in this section. It accounts for a decrease in PV output due to actual conditions such as shading, aging, dust cover, wire losses, and snow. Derating factor in this research study was set at 80%. Figure 7.18a and b demonstrate how the derating factor and capital cost of PV panels affect the profitability of the selected energy solutions. The lowest NPC and LCOE are found when the PV initial cost is lower and the derating factor is higher, respectively. An increase in the capital cost of PV system by 2.5 times in the constant derating factor would increase NPC and LCOE by $4.8k and 0.1/kWh, respectively. Project lifetime also is observed as an important indicator in the economic analysis of renewable HES. Figure 7.19a and b show how LCOE is

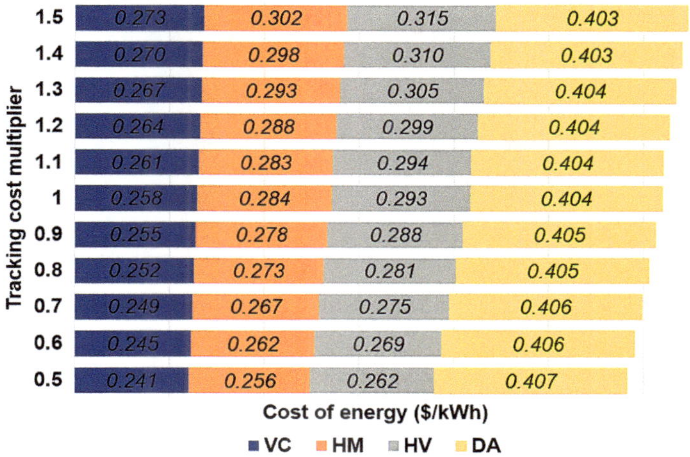

Fig. 7.17 Fluctuation of initial costs of PV trackers against the COE

sensitive to solar irradiation fluctuation in short-term (10 years) and long-term (20 years) project lifetimes. The higher the solar irradiation, the greater PV output can be obtained, resulting in lowering overall energy costs and NPC. The conclusion can be that renewable HES can have more resource values and lower energy costs as the project lifetime increases.

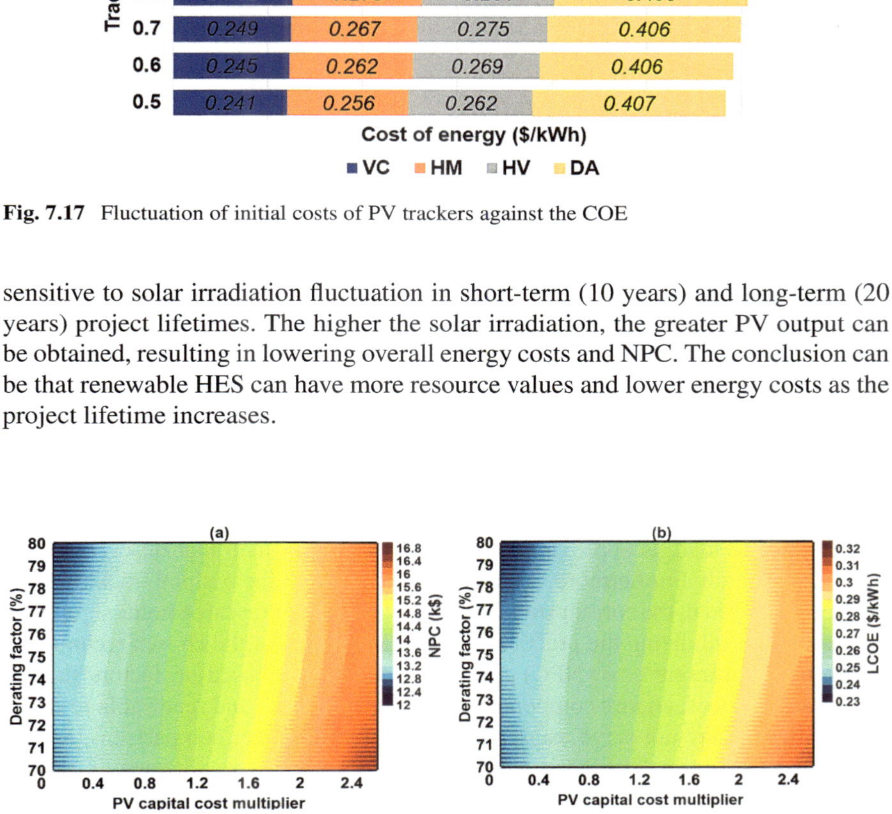

Fig. 7.18 Effect of PV capital cost multiplier and derating factor on **a** NPC and **b** LCOE

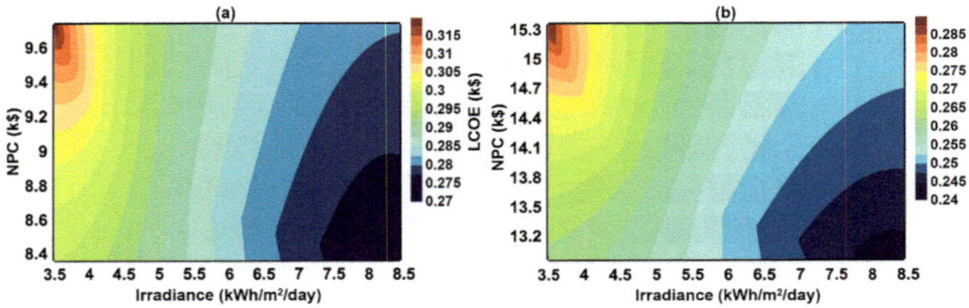

Fig. 7.19 Effect of solar irradiation on LCOE during **a** short-term and **b** long-term project

7.5.4 Comparison & Validation

A comparative analysis between the findings of this present study and previously conducted research in the same domain reveals insights into the promising or unprofitable future of the selected hybrid energy options. However, conducting a direct comparison between the current study and prior works presents challenges due to variations in the hybrid energy system (HES) configuration, technology choices, component sizes, and financial parameters. Nonetheless, the results of the optimal solutions in this study, as depicted in Fig. 7.20, demonstrate a more favorable outcome with a lower net present cost (NPC) and levelized cost of energy (LCOE) by \$58k, and \$0.005/kWh, respectively, compared to the previous research. The earlier study exhibited higher excess electricity, which could potentially be utilized to charge the battery and enhance the reliability of the current optimal solution. Moreover, the analysis in the current study reveals that battery autonomy, defined as the ratio of battery size to electric load, is approximately five times greater than that observed in the previous research. Furthermore, the comparison between the optimal tracking technology employed in the current study and that of the earlier research holds significant importance in validating the practicality and profitability of the chosen technology. While the net present cost (NPC) can vary considerably based on factors such as local tracker prices, energy component costs, required load, and renewable resource availability, it may not serve as a precise indicator for direct comparison. Instead, presenting the optimal trackers and comparing the NPC growth rates of various tracking systems can offer a suitable method to guide consumers toward selecting the most financially and performance-appropriate tracker. Table 7.6 presents the breakdown of the optimal solution and economic parameters of the previous findings which were assessed on different solar tracker technologies. Consistent with other studies akin to the current research, vertical-axis PV trackers (VC) have been identified as favorable solutions due to their cost-effectiveness and reliability. Additionally, the DA tracking technology exhibited higher NPC growth rates across various scenarios.

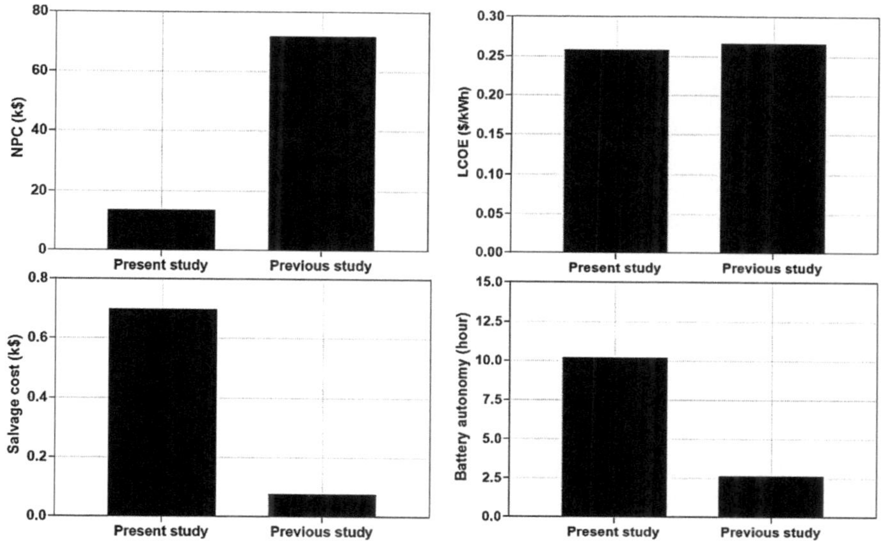

Fig. 7.20 Comparison of the results in the present and previous study

Table 7.6 Comparison of the present results with the literature findings

	Ref., location	HES[1]	Winning tracker[2]	NPC growth rate (%)[3]			
				HM	HC	VC	DT
Grid-connected	[29], Makkah, Saudi Arabia	PV	DA	20.63	20.00	1.00	19.95
	[57], Eight cities in Iran	PV	VC	2.06	3.40	Optimal	0.70
	[39], Alkharj City, Saudi Arabia	PV/ELC/FC	VC	–	78.00	Optimal	84.00
Off-grid	[34], Hilly terrain, India	PV/WT	VC	2.00	3.86	Optimal	9.77
	[58], South of Iran	PV/Pump storage hydroelectric	DA	–	12.9	–	Optimal
	[37], Healthcare facility, India	PV	VC	7.60	16.40	Optimal	13.90
	Kwazulu Natal Province, South Africa	PV/DG/HKT	VC	9.80	13.80	Optimal	18.30

7.6 Conclusion

This study aimed to determine the most stable and profitable hybrid renewable energy alternative using four PV tracking technologies for a residential place in South Africa. The optimal sizing of the hybrid energy solutions consisting of DG, PV system, HKT, battery, and converter was conducted to ascertain the reliable and cost-effective solution. The winning or optimal energy system was selected as per its lowest NPC and LCOE. A combination of DG, PV, and HKT under the VC tracker achieved the most desirable hybrid energy system in terms of cost-effectiveness. The NPC and LCOE for this option, which included 2 kW DG, 3.02 kW PV, 1 HKT, 6 battery units, and 2.12 kW converter, were \$13.7k, \$0.258/kWh, and 281.11 kg/year, respectively. The use of PV/HKT system with VC tracker showed the most appropriate option for electricity production. The highest and lowest CO_2 emissions were observed in HES connected to the FT and DA tracking technologies, respectively. The VC-based DG/PV/HKT option and the HC-based hybrid PV/HKT option are regarded as the most dependable due to their lowest unmet loads. Conversely, DA tackers with both configurations achieve lower reliability to meet the required load. A sensitivity evaluation was performed in the next section to predict how the renewable energy options and PV trackers would change under various technical and economic circumstances. As per the sensitivity analysis on tracking configurations, the VC strategy is more sensitive to the fluctuation of SOC_{min}; rising SOC_{min} will increase NPC and CO_2 by \$0.16k and 1.08 tonnes/year, respectively. Furthermore, VC-based and HM-based hybrid systems show the greatest increase in PV production and RF due to increasing albedo.

7.7 Future Studies

In the future research attempt, a comprehensive life cycle analysis (LCA) can be conducted to assess the environmental impact of solar tracking technologies and the overall renewable energy system over their entire lifecycle. This potential analysis can consider various stages, including manufacturing, installation, operation, and end-of-life. The combination of techno-economic analysis and life cycle cost analysis (LCA) provides a comprehensive assessment of both the financial and environmental aspects of a project. The techno-economic analysis evaluates the financial viability and feasibility by considering capital costs, revenue streams, and payback periods. On the other hand, LCA assesses the total cost of the project over its entire lifecycle, including environmental impacts and energy consumption. Integrating these analyses allows decision-makers to make more informed and balanced choices, considering both economic benefits and environmental implications. This approach leads to the development of sustainable and financially viable solutions that align with both economic and environmental goals, enabling better decision-making and resource allocation.

References

1. J.G. Peña Balderrama et al., Incorporating high-resolution demand and techno-economic optimization to evaluate micro-grids into the Open Source Spatial Electrification Tool (OnSSET). Energy Sustain. Dev. **56**, 98–118 (2020). https://doi.org/10.1016/j.esd.2020.02.009
2. T. Kobayakawa, T.C. Kandpal, A techno-economic optimization of decentralized renewable energy systems: trade-off between financial viability and affordability—a case study of rural India. Energy Sustain. Dev. **23**, 92–98 (2014). https://doi.org/10.1016/j.esd.2014.07.007
3. T. Levin, V.M. Thomas, Can developing countries leapfrog the centralized electrification paradigm? Energy Sustain. Dev. **31**, 97–107 (2016). https://doi.org/10.1016/j.esd.2015.12.005
4. M. Babaei Jamnani, A. Kardgar, Energy-exergy performance assessment with optimization guidance for the components of the 396-MW combined-cycle power plant. Energy Sci. Eng. **8**(10), 3561–3574 (2020). https://doi.org/10.1002/ese3.764
5. T.S. Costa, M.G. Villalva, Technical evaluation of a PV-diesel hybrid system with energy storage: case study in the Tapajós-Arapiuns extractive reserve, Amazon, Brazil. Energies **13**(11), 2969 (2020). https://doi.org/10.3390/en13112969
6. X. Liu, Q. Tan, Y. Niu, R. Babaei, Techno-economic analysis of solar tracker-based hybrid energy systems in a rural residential building: a case study in South Africa. Int. J. Green Energy **20**(2), 192–211 (2023). https://doi.org/10.1080/15435075.2021.2024545
7. J. Lata-García, F. Jurado, L.M. Fernández-Ramírez, H. Sánchez-Sainz, Optimal hydrokinetic turbine location and techno-economic analysis of a hybrid system based on photovoltaic/hydrokinetic/hydrogen/battery. Energy **159**, 611–620 (2018). https://doi.org/10.1016/j.energy.2018.06.183
8. H.J. Vermaak, Techno-economic analysis of solar tracking systems in South Africa. Energy Procedia **61**, 2435–2438 (2014). https://doi.org/10.1016/j.egypro.2014.12.018
9. K. Kusakana, Optimization of the daily operation of a hydrokinetic–diesel hybrid system with pumped hydro storage. Energy Convers. Manag. **106**, 901–910 (2015). https://doi.org/10.1016/j.enconman.2015.10.021
10. K. Kusakana, Optimal operation scheduling of a hydrokinetic-diesel hybrid system with pumped hydro storage, in *2015 4th International Conference on Electric Power and Energy Conversion Systems (EPECS)*, (2015), pp. 1–6. https://doi.org/10.1109/EPECS.2015.7368539
11. M. Shafiey Dehaj, H. Hajabdollahi, Multi-objective optimization of hybrid solar/wind/diesel/battery system for different climates of Iran. Environ. Dev. Sustain. **23**(7), 10910–10936 (2021). https://doi.org/10.1007/s10668-020-01094-1
12. M. Najafi Ashtiani, A. Toopshekan, F. Razi Astaraei, H. Yousefi, A. Maleki, Techno-economic analysis of a grid-connected PV/battery system using the teaching-learning-based optimization algorithm. Sol. Energy **203**, 69–82 (2020). https://doi.org/10.1016/j.solener.2020.04.007
13. S.A. Mousavi, R.A. Zarchi, F.R. Astaraei, R. Ghasempour, F.M. Khaninezhad, Decision-making between renewable energy configurations and grid extension to simultaneously supply electrical power and fresh water in remote villages for five different climate zones. J. Clean. Prod. **279**, 123617 (2021). https://doi.org/10.1016/j.jclepro.2020.123617
14. R. Babaei, D.S. Ting, R. Carriveau, Feasibility and optimal sizing analysis of stand-alone hybrid energy systems coupled with various battery technologies: a case study of Pelee Island. Energy Rep. **8**, 4747–4762 (2022). https://doi.org/10.1016/j.egyr.2022.03.133
15. S. Vendoti, M. Muralidhar, R. Kiranmayi, Techno-economic analysis of off-grid solar/wind/biogas/biomass/fuel cell/battery system for electrification in a cluster of villages by HOMER software. Environ. Dev. Sustain. **23**(1), 351–372 (2021). https://doi.org/10.1007/s10668-019-00583-2
16. A. Shahzad, S. Hanif, Techno-economic feasibility of biogas generation in Attari village, Ferozepur road, Lahore. Environ. Dev. Sustain. **16**(5), 977–993 (2014). https://doi.org/10.1007/s10668-013-9506-5
17. N.R. Deevela, B. Singh, T.C. Kandpal, Techno-economics of solar PV array-based hybrid systems for powering telecom towers. Environ. Dev. Sustain. **23**(11), 17003–17029 (2021). https://doi.org/10.1007/s10668-021-01379-z

18. A.S. Aziz, M.F.N. Tajuddin, M.R. Adzman, M.A.M. Ramli, Impacts of albedo and atmospheric conditions on the efficiency of solar energy: a case study in temperate climate of Choman, Iraq. Environ. Dev. Sustain. **23**(1), 989–1018 (2021). https://doi.org/10.1007/s10668-019-00568-1
19. T. Chen, M. Wang, R. Babaei, M.E. Safa, A.A. Shojaei, Technoeconomic analysis and optimization of hybrid solar-wind-hydrodiesel renewable energy systems using two dispatch strategies. Int. J. Photoenergy **2023** (2023). https://doi.org/10.1155/2023/3101876
20. A. Amirsoleymani, R. Babaei, S.S. Mousavi Ajarostaghi, M. Saffari Pour, Feasibility evaluation of stand-alone energy solutions in energy-poor Islands using sustainable hydrogen production. Int. J. Energy Res. **46**(15), 24045–24063 (2022). https://doi.org/10.1002/er.8704
21. M.A. Mohamed, H.M. Abdullah, M.A. El-Meligy, M. Sharaf, A.T. Soliman, A. Hajjiah, A novel fuzzy cloud stochastic framework for energy management of renewable microgrids based on maximum deployment of electric vehicles. Int. J. Electr. Power Energy Syst. **129**, 106845 (2021). https://doi.org/10.1016/j.ijepes.2021.106845
22. M.A. Mohamed, T. Jin, W. Su, Multi-agent energy management of smart islands using primal-dual method of multipliers. Energy **208**, 118306 (2020). https://doi.org/10.1016/j.energy.2020.118306
23. H. Zou, J. Tao, S.K. Elsayed, E.E. Elattar, A. Almalaq, M.A. Mohamed, Stochastic multi-carrier energy management in the smart islands using reinforcement learning and unscented transform. Int. J. Electr. Power Energy Syst. **130**, 106988 (2021). https://doi.org/10.1016/j.ijepes.2021.106988
24. B. Bhandari, K.-T. Lee, C.S. Lee, C.-K. Song, R.K. Maskey, S.-H. Ahn, A novel off-grid hybrid power system comprised of solar photovoltaic, wind, and hydro energy sources. Appl. Energy **133**, 236–242 (2014). https://doi.org/10.1016/j.apenergy.2014.07.033
25. F.K. Abo-Elyousr, A. Elnozahy, Bi-objective economic feasibility of hybrid micro-grid systems with multiple fuel options for islanded areas in Egypt. Renew. Energy **128**, 37–56 (2018). https://doi.org/10.1016/j.renene.2018.05.066
26. M.K. Shahzad, A. Zahid, T. ur Rashid, M.A. Rehan, M. Ali, M. Ahmad, Techno-economic feasibility analysis of a solar-biomass off grid system for the electrification of remote rural areas in Pakistan using HOMER software. Renew. Energy **106**, 264–273 (2017). https://doi.org/10.1016/j.renene.2017.01.033
27. A. Oulis Rousis, D. Tzelepis, I. Konstantelos, C. Booth, G. Strbac, Design of a hybrid AC/DC microgrid using HOMER Pro: case study on an islanded residential application. Inventions **3**(3), 55 (2018). https://doi.org/10.3390/inventions3030055
28. M.H. Jahangir, S.A. Mousavi, R. Asayesh Zarchi, Implementing single- and multi-year sensitivity analyses to propose several feasible solutions for meeting the electricity demand in large-scale tourism sectors applying renewable systems. Environ. Dev. Sustain. **23**(10), 14494–14527 (2021). https://doi.org/10.1007/s10668-021-01254-x
29. H.Z. Al Garni, A. Awasthi, M.A.M. Ramli, Optimal design and analysis of grid-connected photovoltaic under different tracking systems using HOMER. Energy Convers. Manag. **155**, 42–57 (2018). https://doi.org/10.1016/j.enconman.2017.10.090
30. V. Boddapati, A.S.R. Nandikatti, S.A. Daniel, Techno-economic performance assessment and the effect of power evacuation curtailment of a 50 MWp grid-interactive solar power park. Energy Sustain. Dev. **62**, 16–28 (2021). https://doi.org/10.1016/j.esd.2021.03.005
31. R. Srivastava, A.N. Tiwari, V.K. Giri, An overview on performance of PV plants commissioned at different places in the world. Energy Sustain. Dev. **54**, 51–59 (2020). https://doi.org/10.1016/j.esd.2019.10.004
32. M.H.M. Sidek, N. Azis, W.Z.W. Hasan, M.Z.A. Ab Kadir, S. Shafie, M.A.M. Radzi, Automated positioning dual-axis solar tracking system with precision elevation and azimuth angle control. Energy **124**, 160–170 (2017). https://doi.org/10.1016/j.energy.2017.02.001
33. V. Sumathi, R. Jayapragash, A. Bakshi, P. Kumar Akella, Solar tracking methods to maximize PV system output—a review of the methods adopted in recent decade. Renew. Sustain. Energy Rev. **74**, 130–138 (2017). https://doi.org/10.1016/j.rser.2017.02.013
34. S. Sinha, S.S. Chandel, Analysis of fixed tilt and sun tracking photovoltaic–micro wind based hybrid power systems. Energy Convers. Manag. **115**, 265–275 (2016). https://doi.org/10.1016/j.enconman.2016.02.056

35. S. Bhakta, V. Mukherjee, Techno-economic viability analysis of fixed-tilt and two axis tracking stand-alone photovoltaic power system for Indian bio-climatic classification zones. J. Renew. Sustain. Energy **9**(1), 015902 (2017). https://doi.org/10.1063/1.4976119
36. C. Li, W. Yu, Techno-economic comparative analysis of off-grid hybrid photovoltaic/diesel/battery and photovoltaic/battery power systems for a household in Urumqi, China. J. Clean. Prod. **124**, 258–265 (2016). https://doi.org/10.1016/j.jclepro.2016.03.002
37. N.M. Kumar, J. Vishnupriyan, P. Sundaramoorthi, Techno-economic optimization and real-time comparison of sun tracking photovoltaic system for rural healthcare building. J. Renew. Sustain. Energy **11**(1), 015301 (2019). https://doi.org/10.1063/1.5065366
38. H. Al Garni, A. Awasthi, Techno-economic feasibility analysis of a solar PV grid-connected system with different tracking using HOMER software, in *2017 IEEE International Conference on Smart Energy Grid Engineering (SEGE)* (2017), pp. 217–222. https://doi.org/10.1109/SEGE.2017.8052801
39. S. Mubaarak et al., Techno-economic analysis of grid-connected PV and fuel cell hybrid system using different PV tracking techniques. Appl. Sci. **10**(23), 8515 (2020). https://doi.org/10.3390/app10238515
40. X. Liu et al., Techno-economic analysis of solar tracker-based hybrid energy systems in a rural residential building: a case study in South Africa. Int. J. Green Energy **00**(00), 1–20 (2022). https://doi.org/10.1080/15435075.2021.2024545
41. M.R. Elkadeem, S. Wang, S.W. Sharshir, E.G. Atia, Feasibility analysis and techno-economic design of grid-isolated hybrid renewable energy system for electrification of agriculture and irrigation area: a case study in Dongola, Sudan. Energy Convers. Manag. **196**, 1453–1478 (2019). https://doi.org/10.1016/j.enconman.2019.06.085
42. K. Kusakana, Techno-economic analysis of off-grid hydrokinetic-based hybrid energy systems for onshore/remote area in South Africa. Energy **68**, 947–957 (2014). https://doi.org/10.1016/j.energy.2014.01.100
43. Resources, NASA Prediction of Worldwide Energy Resources (2021). https://power.larc.nasa.gov
44. K. Kusakana, H.J. Vermaak, Hydrokinetic power generation for rural electricity supply: case of South Africa. Renew. Energy **55**, 467–473 (2013). https://doi.org/10.1016/j.renene.2012.12.051
45. R. Babaei, D.S.-K. Ting, R. Carriveau, Feasibility and optimal sizing analysis of stand-alone hybrid energy systems coupled with various battery technologies: a case study of Pelee Island. Energy Rep. **8**, 4747–4762 (2022). https://doi.org/10.1016/j.egyr.2022.03.133
46. 1KW 2KW 3KW 5KW Windmill Generator System/2KW 3KW 5KW Wind Turbine Kit, "Alibaba." https://www.alibaba.com/product-detail/1KW-2KW-3KW-5KW-Windmill-Generator_60493711744.html?spm=a2700.details.0.0.5bce10144u6yxW
47. ¢/KWH Levelized Cost of Electricity for Various Power and Energy Efficiency Options, National Hydropower Association. https://www.hydro.org/waterpower/why-hydro/affordable
48. C.S. Lai et al., Levelized cost of electricity for photovoltaic/biogas power plant hybrid system with electrical energy storage degradation costs. Energy Convers. Manag. **153**, 34–47 (2017). https://doi.org/10.1016/j.enconman.2017.09.076
49. M.R. Akhtari, M. Baneshi, Techno-economic assessment and optimization of a hybrid renewable co-supply of electricity, heat and hydrogen system to enhance performance by recovering excess electricity for a large energy consumer. Energy Convers. Manag. **188**, 131–141 (2019). https://doi.org/10.1016/j.enconman.2019.03.067
50. T. Salameh, M.A. Abdelkareem, A.G. Olabi, E.T. Sayed, M. Al-Chaderchi, H. Rezk, Integrated standalone hybrid solar PV, fuel cell and diesel generator power system for battery or supercapacitor storage systems in Khorfakkan, United Arab Emirates. Int. J. Hydrogen Energy **46**(8), 6014–6027 (2021). https://doi.org/10.1016/j.ijhydene.2020.08.153
51. M.H. Jahangir, S.A. Mousavi, M.A. Vaziri Rad, A techno-economic comparison of a photovoltaic/thermal organic Rankine cycle with several renewable hybrid systems for a residential area in Rayen, Iran. Energy Convers. Manag. **195**, 244–261 (2019). https://doi.org/10.1016/j.enconman.2019.05.010

52. T. Aderinto, H. Li, Ocean wave energy converters: status and challenges. Energies **11**(5), 1250 (2018). https://doi.org/10.3390/en11051250
53. M. Baneshi, F. Hadianfard, Techno-economic feasibility of hybrid diesel/PV/wind/battery electricity generation systems for non-residential large electricity consumers under southern Iran climate conditions. Energy Convers. Manag. **127**, 233–244 (2016). https://doi.org/10.1016/j.enconman.2016.09.008
54. S. Mandal, B.K. Das, N. Hoque, Optimum sizing of a stand-alone hybrid energy system for rural electrification in Bangladesh. J. Clean. Prod. **200**, 12–27 (2018). https://doi.org/10.1016/j.jclepro.2018.07.257
55. F. Fazelpour, N. Soltani, M.A. Rosen, Economic analysis of standalone hybrid energy systems for application in Tehran, Iran. Int. J. Hydrogen Energy **41**(19), 7732–7743 (2016). https://doi.org/10.1016/j.ijhydene.2016.01.113
56. C. Bordin, H.O. Anuta, A. Crossland, I.L. Gutierrez, C.J. Dent, D. Vigo, A linear programming approach for battery degradation analysis and optimization in offgrid power systems with solar energy integration. Renew. Energy **101**, 417–430 (2017). https://doi.org/10.1016/j.renene.2016.08.066
57. M.A. Vaziri Rad, A. Toopshekan, P. Rahdan, A. Kasaeian, O. Mahian, A comprehensive study of techno-economic and environmental features of different solar tracking systems for residential photovoltaic installations. Renew. Sustain. Energy Rev. **129**, 109923 (2020). https://doi.org/10.1016/j.rser.2020.109923
58. M. Shabani, J. Mahmoudimehr, Techno-economic role of PV tracking technology in a hybrid PV-hydroelectric standalone power system. Appl. Energy **212**, 84–108 (2018). https://doi.org/10.1016/j.apenergy.2017.12.030

Chapter 8
Metal Organic Frameworks (MOFs) in Adsorption Heat Transformations

Joud Al-rabadi and Julia Aman

Abstract This report reviews the performance of metal–organic frameworks (MOFs) in Adsorption Heat Transformation (AHT) systems. MOFs are porous materials characterized with high surface areas and tunable properties that make them suitable for a range of applications. This study concentrates on the use of MIL-101(Cr) in a 10-kW cooling capacity system and compares its performance with activated carbon. Results show that less mass of MIL-101(Cr) is required to produce 10 kW of power compared to activated carbon, while also having a higher coefficient of performance. The COP of MIL-101(Cr) is calculated to be about 1.5 times higher than that of activated carbon, highlighting the potential of MOFs in refrigeration applications. However, MOFs are still in the preliminary stages of development and require more research to overcome challenges and prove their commercial viability.

Nomenclature

COP	Coefficient of performance
C_p	Specific heat at constant pressure (kJ/kg-K)
h	Specific enthalpy (kJ/Kg)
$\dot{m}$	Mass flow rate (kg/s)
P	Pressure (kPa)
$\dot{Q}$	Heat transfer rate (kW)
q_{sh}	Heat of sorption (kJ/kg)
R	Universal gas constant (J/kg-K)
t_{cycle}	Cycle time (hr)
T	Temperature (K)
T_A	Adsorption Temperature (K)
T_C	Desorption Temperature (K)

J. Al-rabadi · J. Aman (✉)
Department of Mechanical, Industrial and Aerospace Engineering, Concordia University, 1455 De Maisonneuve Blvd. W., Montreal, QC H3G 1M8, Canada
e-mail: julia.aman@concordia.ca

D. S-K. Ting and A. Vasel-Be-Hagh (eds.), *Engineering to Adapt*, Springer Proceedings in Energy, https://doi.org/10.1007/978-3-031-47237-4_8

X	Mass fraction of ammonia (%)
x	Ammonia concentration (kg NH_3/kg carbon)

Subscripts

ads	Adsorber
con or *H*	Condenser
des	Desorber
eva or *L*	Evaporator
in	Inlet
max	Maximum
MOF	Metal Organic Framework
min	Minimum
out	Outlet
ref	Refrigerant
sat	Saturation
sol	Solution

8.1 Introduction

Over the last two decades, prioritizing the use of renewable energy sources and implementing more energy-efficient processes have been of utmost importance. To ensure sustainability in the future, researchers are now focusing on finding more environmentally friendly solutions. Among various other crucial processes, the refrigeration process relies heavily on fossil fuels and accounts for approximately 25% of the total global energy production [1]. The dependence on fossil fuels could be substantially reduced by adopting reliable and efficient alternatives such as solar and wind energy, which would also eliminate harmful chemical emissions. However, the development of new technologies requires extensive research to reduce the cost and increase the chances of its implementation to compete with conventional methods. The refrigeration cycle comprises several modules, offering various options for modifications to enhance its sustainability. One potential approach is to identify alternatives to the compressor, which typically requires a significant amount of electrical energy to operate.

Traditional chillers and heat-pumps rely on vapor compression cycle which is driven by electrical energy. However, recent advancements have introduced a more environmentally friendly alternative called adsorption-desorption cycle. Research shows that the adsorption-desorption cycle requires low-grade heat to initiate, which can be obtained from solar energy or more desirably, residual heat from other processes [2]. In addition, the adsorption-desorption cycle performs the dual function of cooling and heating, negating the need for separate units, as opposed to compression systems where two different units are needed.

The vapor compression cycle operates as a closed system using a refrigerant as the working fluid (Fig. 8.1). During this process, the saturated refrigerant vapor undergoes compression by the compressor, resulting in superheated refrigerant vapor with an increase in pressure and temperature, which enables the refrigerant to be condensed into a 2-phase mixture upon passing through the condenser. The condenser is characterized by a heat release to the surroundings from the high temperature superheated refrigerant vapor, resulting in saturated/subcooled liquid mixture. Afterward, the pressure and temperature of the mixture are reduced through the use of an expansion valve. The low pressure and temperature mixture flows towards the evaporator where it gains energy from the surrounding, transforming it back into a saturated vapor state. The evaporator's temperature is maintained at a relatively low level due to the phase change process [2].

The adsorption process is classified under the sorption category which also includes the absorption process. Absorption occurs due to a phenomenon called diffusion which happens when the particles of a substance, known as the absorbate, are taken in by another material, referred to as the absorbent [4]. On the other hand, adsorption is a reversible process that takes place when the particles of a fluid, called the adsorbate, adhere to the surface of a porous solid matrix, referred to as the adsorbent [5]. This results in the formation of a monomolecular layer on top of the adsorbent. silica gel and zeolites are commonly used adsorbents in the industry;

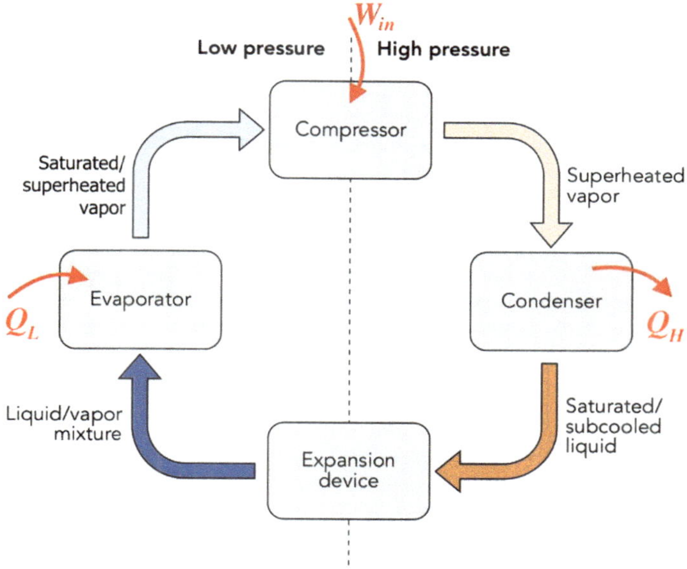

Fig. 8.1 The conventional vapor compression cycle [3]

however, MOFs have shown better potential [6]. In general, adsorption technologies present a significant implementation challenge due to concerns surrounding the regeneration, reusability, and stability of such systems. Additionally, performance, unit size, and production costs are often higher than conventional systems, especially since, thus far, adsorption systems have not demonstrated significant efficiency gains relative to existing systems [7].

8.2 What Are Metal–Organic Frameworks?

MOFs comprise organic linkages and metal cores as shown in Fig. 8.2. Thanks to their crystalline structure, they are considered an advanced class of porous adsorbents and unique in the sense that different structures can be synthesized according to the respective application of the MOF, therefore, MOFs have a wide scope of applications, and their efficiency can be increased by tuning their structure [8].

Recently MOFs have gained significant attention due to their distinctive properties, such as their ability to adjust pore size, low generation temperatures, and rapid adsorption-desorption dynamics, among others. The structure of MOF allows tuned synthesis to achieve certain properties for specific applications. Variability in the surface of the MOF such as defects or different crystallinity can result in various properties for heat transformation.

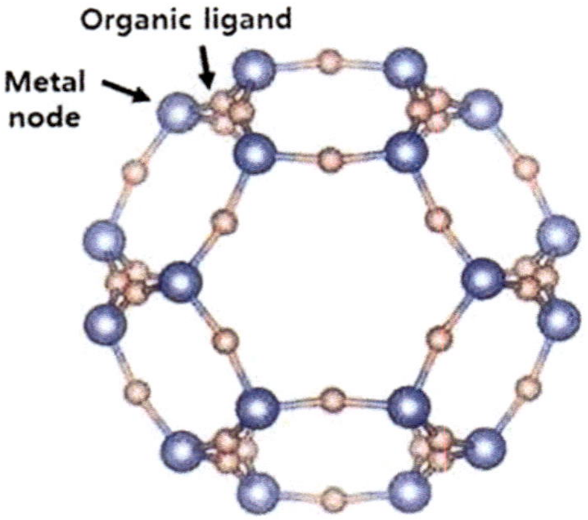

Fig. 8.2 The structure and the building blocks of MOFs [9]

8.3 Applications of MOFs

The properties of MOFs described earlier are highly suitable for use as adsorbent materials. This paper focuses on the applications of MOFs as an adsorbent in heat transformation systems, however, there are other applications for MOFs where they act as adsorbents as well, namely: water harvesting, gas storage, carbon capture, and adsorption heat storage.

8.3.1 Water Harvesting

Many regions in the world face water scarcity issues; water harvesting is a potential solution to this problem. This process involves extracting water molecules from the air and storing it. This can be made possible by the use of MOFs since their structure can be tuned to selectively adsorb water molecules from the air. In addition, MOFs have demonstrated the ability to be regenerated at certain temperatures, rendering the method reusable [6]. Extensive research must be done to ensure the final structure synthesized provides high moisture capture capacity and relatively easy regeneration of MOFs [10].

8.3.2 Gas Storage

Renewable alternatives to fossil fuels include gases such as Hydrogen and Methane; however, their practicality is limited by the need for large storage volumes. Nonetheless, MOFs have made achieving both high-capacity uptake and high deliverable capacity possible, making these gases more viable options. By having a distinguishably high specific surface area, MOFs can store more gas than other materials [6].

8.3.3 Carbon Capture

Separation processes play a crucial role in numerous applications. Despite the advancements, several obstacles still need to be overcome including safety concerns, high energy consumption, and complex equipment. Fortunately, MOFs have emerged as potential alternatives to the current conventional methods since the main technology remains in the MOFs material itself. Once the structure of a suitable MOF for a specific chemical requiring separation is determined, the process is relatively simple to implement and cost-effective. Thanks to MOFs properties, such as high selectivity and a large specific surface area, they are able to outperform other methods [10].

8.3.4 Adsorption Heat Storage

There are three different types of thermal energy storage: sensible, latent, and thermo-chemical. Adsorption heat storage belongs to the thermo-chemical category, and it needs two different sources or sinks to function making it an indirect thermal energy storage process. The efficiency of the system is reliant on the adsorbent material and density. For instance, MOFs exhibit a high uptake capacity due to their crystalline pore structure. Since heat is required to initiate the desorption process, this results in adsorption potential energy where the energy is trapped within the system until the adsorbate makes contact with the adsorbent. This represents one-half of the adsorption heat transformation cycle. The energy can be stored until the adsorbate gains access to the adsorbent again where the discharging phase commences, resulting in the production of heat. This application can be used when the refrigeration or heat-pumping unit is needed at night after it is charged during the day for instance [11].

8.4 Adsorption Heat Transformation System

Adsorption Heat Transformation (AHT) is a process that can be driven by green-energy sources, which can replace the electrically driven compressor in the conventional vapor compression cycle in refrigeration and heat-pump systems. It involves two theoretically reversible processes, namely adsorption and desorption. Adsorption is a spontaneous process where the adsorbate, a fluid, adheres to the surface of the adsorbent. During this process, the molecules of the adsorbate get more stable leading to the evolution of heat: therefore, it is considered as an exothermic process. While desorption is endothermic since the adsorbate needs to obtain enough energy to exceed the activation barrier that keeps the adsorbate bonded to the surface of the adsorbent.

AHT systems can either be open-cycle or closed-cycle. The key difference between the two is that in the former the mass entering the system is not necessarily equal to the mass leaving the system, while in the latter, there is no change in mass [7]. One of the distinctive features of AHT is that the heat released during the adsorption and condensation phase can be used to heat up the liquid to turn it into vapor in the evaporator, which creates a regenerative cycle.

AHT comprises two primary compartments for two cycles: the working cycle and the regeneration cycle (Fig. 8.3). In the latter, desorption occurs after heat is derived from a low-grade heat source and then the desorbed vapor flows to the condenser where heat is released (heating). The regenerative cycle switches to the

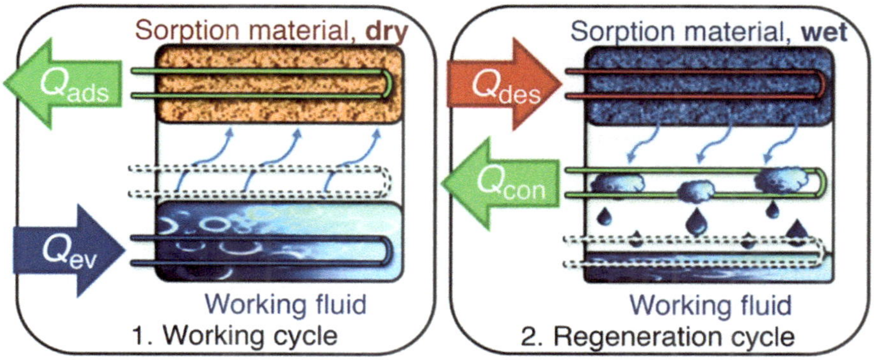

Fig. 8.3 Working cycle and regeneration cycles of an adsorption system [7]

working cycle once the quantity of adsorbed fluid meets specific requirements. The evaporator exploits the required heat from its surroundings, which results in a drop in temperature in the surrounding (cooling) [7].

8.5 Potential of MOFs in Heat Adsorption Systems

As previously mentioned in the paper, MOFs possess distinctive properties that make them suitable materials for serving as adsorbents in adsorption systems. The utilization of MOFs in AHT has been studied and its implementation is dependent on the water stability of the MOF which is affected by the reversible nature of the bonds in the molecule. Scientific research is underway to synthesize and study different MOFs with the aim of enhancing their structural stability in the presence of water [8].

The most used closed-cycle system-based applications of MOFs in the heat transfer process are adsorption heat pumps and chillers [12]. However, since high hydrothermal stability is required for MOFs to function effectively as adsorbents, there exist very few MOFs that are suitable for AHT systems. MIL-101(Cr), Aluminum fumarate, and CAU-10-H among others are potential adsorbents with high uptake capacity and hydrothermal stability [7].

MOFs characteristics allow them to outperform other traditionally used adsorbents such as silica gel and zeolites in AHT applications; MOFs produce fewer Carbon dioxide (CO_2) emissions and have higher efficiency and water uptake [6]. For instance, Zeolites exhibit high hydrophilicity which results in the need for temperatures as high as 120 °C to initiate desorption processes [7].

Silica gel has low efficiency due to its tendency to adsorb water at higher partial pressures. Therefore, it is challenging to achieve high levels of saturation at standard surrounding conditions given a fixed quantity of Silica gel. Moreover, it exhibits low water uptake capacity which affects the working fluid flow rate and the size of the adsorption system. MOFs have lower partial pressure and higher water uptake capacity, rendering them more desirable adsorbents [12].

8.6 Adsorption Heat Transformation (AHT) Systems for Heating and Cooling Applications

Adsorption heat pumps (AHP) and air-conditioning systems have many advantages over conventional heating and cooling systems. In adsorption heating and cooling systems, the electrically driven compression chiller in the conventional system will be replaced by a thermally driven one. The thermal energy from green-energy resources or waste heat can power the desorption process [13]. In AHT heating and cooling systems, water, an environmentally friendly refrigerant, will replace the toxic refrigerants that are used in conventional systems. Furthermore, less maintenance is required to ensure the functionality of the system since no moving parts are needed [14].

8.7 10 kW MOFs-Based Adsorption Air-Conditioning System

A model is developed of a 10 kW MOFs-based adsorption air-conditioning system following the operating conditions mentioned in Table 8.1 [15]. In this study, two working pairs are compared against each other: MIL-101(Cr)—NH_3 and activated carbon—NH_3. The former working pair is investigated in this paper while the results for the latter were obtained by the authors P. Henshaw, J. Aman et al. Both systems are considered at the same operating conditions.

Table 8.1 Operating conditions for an adsorption cooling system

Desorber T_{des}	80	°C
Condenser T_{con}	30	°C
Evaporator T_{eva}	2	°C
Adsorber T_{ads}	30	°C
Adsorber exit pressure	462	kPa
Desorber exit pressure	1167	kPa
Required mass flow rate of Ammonia	0.0089	kg/s

8.7.1 Selecting the Ideal MOF for Adsorption: Factors Influencing Morphology and Properties

As stated earlier, the MOFs' morphology changes according to the metal ions and organic ligands composing it as well as the production method used. The choice of MOF for adsorption purposes is directly linked to its unique features. The selection process is significant, since the properties of the MOF, such as sorption stability, adsorption capacity, absence of sorption hysteresis, and surface characteristics, have a significant impact on achieving a highly efficient and cost- effective adsorption-desorption system [16].

The organic component in the MOF can consist of either carboxylates or anions such as phosphonate, sulfonate, and heterocyclic compounds. On the other hand, the inorganic component can be either a metal ion or a cluster called Secondary Building Units (SBUs). The MOF's morphology depends on the coordination number, coordination geometry of the metal ions and the characteristics of the functional groups involved [17].

8.7.2 MIL-101(Cr)-Ammonia Solid Sorption Working Pair

The adsorption performance of different MOFs, namely MIL-101(Cr), ZIF-8, and UiO-66 is investigated with various gases including N_2, CH_4, and CO_2. The results indicated that MIL-101(Cr) had the highest adsorption performance compared to the rest. Figure 8.4 presents the data.

Therefore MIL-101(Cr) is chosen in this study as the adsorbent. Additionally, MIL-101(Cr) proved to have high ammonia sorption stability, relatively high sorption capacity compared to other MOFs and insignificant hysteresis. The MIL-101(Cr)-ammonia working pair has 3 times more sorption capacity compared to MIL-101(Cr)-water working pair when the evaporation temperature is less than 8.4 °C. Ammonia is the adsorbate to be studied in this analysis. It is important to prevent the contamination of ammonia with other substances such as water, since the MOF might favor the contaminant over Ammonia resulting in decreased levels of adsorption for ammonia [19].

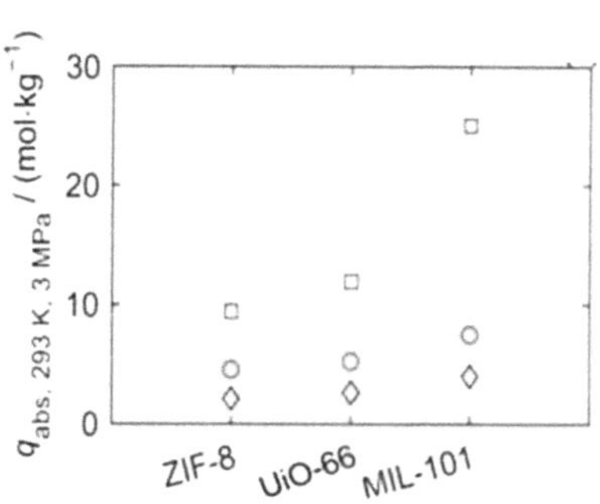

Fig. 8.4 Absolute adsorption capacity (q_{abs}) showing MIL-101 has 3 times more capacity than UiO-66 with □ CO_2, *pink* ○ CH_4, and *blue*◇ N_2 [18]

8.7.3 Adsorption Cycle

The adsorption-desorption refrigeration cycle, shown in Fig. 8.5, is initiated by supplying low-grade heat, such as solar energy, to the adsorption bed that contains a maximum concentration, x_{max} in Fig. 8.6, of adsorbent. The desorption process causes the refrigerant to become a vapor, increasing its pressure and temperature, and causing it to flow towards the condenser once the pressure of the refrigerant matches the condenser's pressure. The refrigerant is condensed by rejecting heat to the lower surrounding temperature until the minimum concentration, x_{min} in Fig. 8.6, is reached. Once this occurs, the heating stops, and the adsorption bed returns to its starting temperature. The refrigerant passes through the throttle valve to reduce its pressure to the evaporator's pressure, where it absorbs heat from the surroundings and vaporizes. The refrigerant then flows back to the adsorption bed, where it is adsorbed by the MOF, completing the cycle.

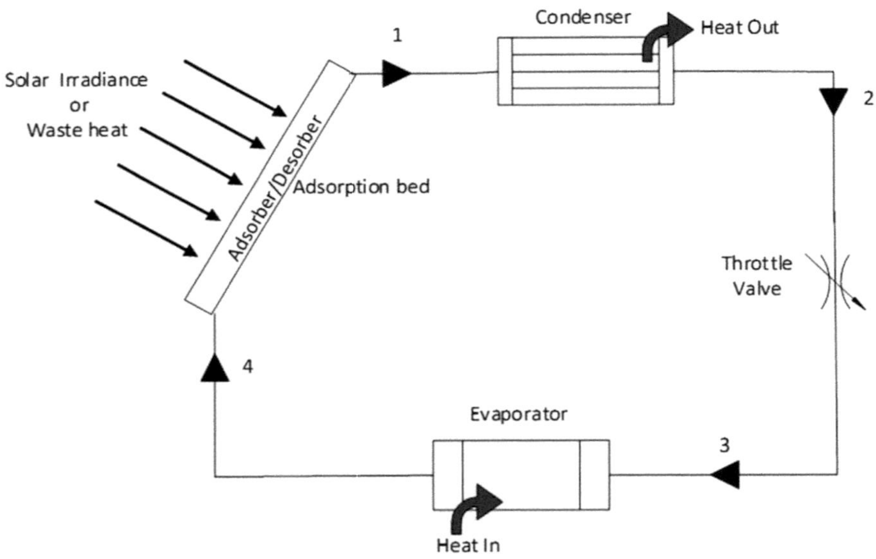

Fig. 8.5 Schematic diagram of an adsorption system

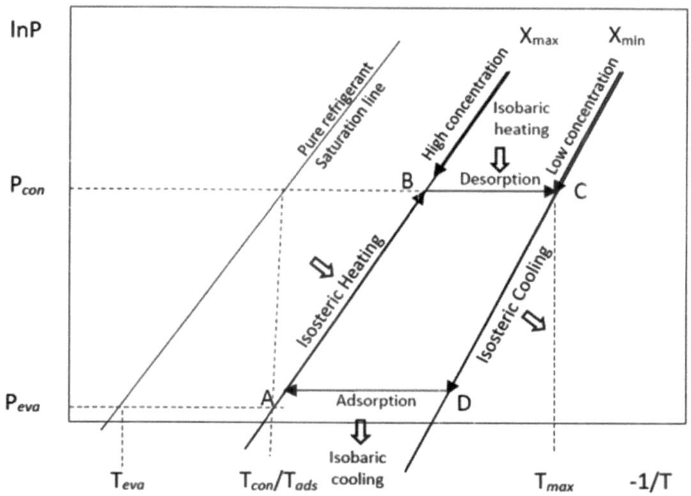

Fig. 8.6 Clapeyron diagram illustrating the thermodynamic cycle of an adsorption cooling system [15]

8.7.4 Analytical Method

Certain assumptions were made to develop the energy flow equations required to obtain the COP of the cooling system, including:

- No heat loss in the system resulting in the temperature at the end of adsorption to be equal to the condensing temperature.
- Adsorption and desorption cycles involve pure NH_3.
- The temperature in the adsorbent bed is consistent, and the refrigerant is adsorbed uniformly throughout the entire bed.
- The specific heat of the refrigerant in the adsorbed phase remains constant, equal to the gas at a given pressure and temperature.
- Heat source and sink temperatures remain constant.
- The isosteric heat of adsorption remains constant.
- The rate of desorption is equivalent to the rate of absorption.
- The full cycle operates for two hours, consisting of one hour of desorption and one hour of adsorption.

To calculate the efficiency of the system, the required heat to initiate desorption is a critical value to find. Equation (8.1) is a combination of the two processes in the cycle which absorb heat, desorption (isosteric) and evaporation (isobaric). The first two terms refer to the heat adsorbed by MOF adsorption bed when the temperature increases from T_A to T_C while the last term represents the desorption heat needed for the x to drop by a specific delta [15].

$$\dot{Q}_{des} = \dot{Q}_{in} = \left[(mC_p)_{MOF}(T_C - T_A) + m_{MOF}x_{\max}C_{pNH_3}(T_C - T_A) + m_{MOF}q_{sh}\Delta x\right]/t_{cycle} \tag{8.1}$$

The specific heat capacity of MIL-101(Cr) can be assumed to be 1.00 kJ/Kg. K according to the [16]. And the heat of sorption, q_{sh}, can be calculated using Eq. (8.2) such that R is equal to the gas constant, A is the slope of the saturation curve on a ln(P) against $\frac{-1}{T_{sat}}$ graph, and T_{sat} is the saturation temperature at the gas pressure P [15].

$$q_{sh} = RA\frac{T}{T_{sat}} \tag{8.2}$$

On the other hand, Eq. (8.3) is used to obtain the rate of heat produced during adsorption as shown below.

$$\dot{Q}_{ads} = \dot{Q}_{out} = \left[(mC_p)_{MOF}(T_C - T_A) + m_{MOF}x_{\min}C_{pNH_3}(T_C - T_A) + m_{MOF}q_{sh}\Delta x + m_{MOF}\Delta x C_{pNH_3}(T_{eva} - T_{ads})\right]/t_{cycle} \tag{8.3}$$

Figure 8.7 shows the ammonia sorption capacity of MIL-101(Cr); at 80 °C the sorption capacity is approximately 0.265 kg of ammonia per 1 kg of MIL-101(Cr). Extrapolation was performed to obtain the maximum sorption capacity at 4.62 bar 30 °C using Fig. 8.7a and b to get 0.65 kg of ammonia per 1 kg of MIL-101(Cr).

The condenser's and evaporator's energy balance formulae are shown below.

$$\dot{Q}_{\text{con}} = \dot{m}_1(h_1 - h_2) \tag{8.4}$$

$$\dot{Q}_{\text{eva}} = \dot{m}_1(h_4 - h_3) \tag{8.5}$$

Equation (8.6) is used to calculate the coefficient of performance of the adsorption chiller.

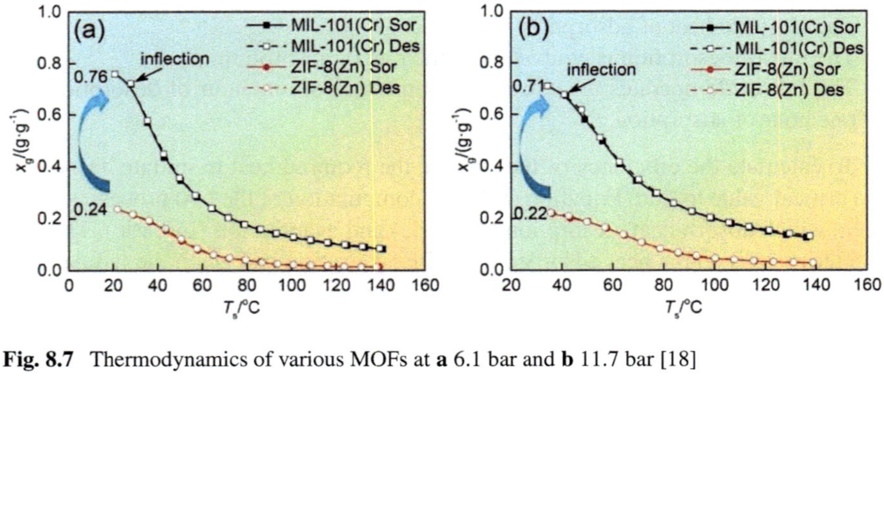

Fig. 8.7 Thermodynamics of various MOFs at **a** 6.1 bar and **b** 11.7 bar [18]

$$COP = \frac{Useful\ output\ energy}{Primary\ energy\ supplied} = \frac{\dot{Q}_{eva}}{\dot{Q}_{des}} \tag{8.6}$$

Equation (8.7) is used to calculate the Carnot Efficiency of the adsorption chiller.

$$Carnot\ Efficiency = 1 - \frac{T_L}{T_H} \tag{8.7}$$

8.7.5 Results

Using Table 8.1 as well as other thermophysical properties of ammonia and MIL-101(Cr), the following was calculated for a total of two hours long cycle, the refrigerant's mass flowrate is 0.0089 kg/s, heat capacity of Ammonia at 30 °C is equal to 2.44 kJ/kg K at the evaporator's pressure of 4.62 bar and 2.59 kJ/kg K at the condenser's pressure of 11.67 bar. The sorption capacity was calculated using Eq. (8.2), the energy flows were calculated using Eqs. (8.1), (8.3), and (8.4) while $\dot{Q}_{eva}$ is given as 10 kW. Using the obtained values, the coefficient of performance was calculated using Eq. (8.6). The results are summarized in Table 8.2.

The COP achieved by this adsorption refrigeration system is found to be lower than that of conventional compression- based refrigeration systems, mainly because of the difference in their energy conversion processes. While conventional refrigeration systems employ electrical energy to initiate the refrigeration cycle, resulting in high COP, MOFs based adsorption refrigeration system utilizes renewable thermal energy, resulting in a relatively lower COP. However, the use of renewable energy sources in system makes it an environmentally friendly alternative to conventional refrigeration systems.

Table 8.2 Summary of the energy flow for different components

x_{max}	0.65	(kg/kg adsorbent)
x_{min}	0.27	(kg/kg adsorbent)
q_{sh}	1377.82	kJ/kg
$\dot{Q}_{con}$	11.45	kW
$\dot{Q}_{eva}$	10	kW
$\dot{Q}_{des}$	17.36	kW
$\dot{Q}_{ads}$	15.56	kW
COP	0.58	-
Carnot COP	22	

Table 8.3 COP, mass of adsorbent required, and Q_{des} of different adsorbents for 10 kW cooling capacity

	MIL-101(Cr)	Activated carbon [16]
COP	0.58	0.35
Mass of adsorbent required (kg)	83.12	671.35
$\dot{Q}_{des}$ (kW)	17.36	28.68

8.7.6 Comparing Coefficient of Performance: MIL-101(Cr) Versus Activated Carbon

MOFs are a new emerging technology in comparison to the traditional activated carbon. Consequently, having a baseline comparison is crucial to explain the advantages and disadvantages associated with MOF implementation. This comparative analysis is of importance due to the similar characteristics of both materials, including: their porous nature and comparable adsorption capacities. On the other hand, MOFs are more selective targeting specific substances effectively while activated carbon's strengths lie in its superior stability and less complex manufacturing process.

Table 8.3 presents a comparison between a 10-kW adsorption refrigeration system that utilizes activated carbon and MOFs as adsorbent and ammonia as the refrigerant. It should be noted that the surface area of Metal-Organic Frameworks (MOFs) is roughly four times greater than that of activated carbon.

The results indicate that the efficiency of the MOF in adsorption–desorption process surpasses that of activated carbon. In addition, the mass of MOF required is significantly less compared to activated carbon making it a more cost-effective option. Moreover, higher rate of heat transfer is required to initiate the activated carbon adsorption–desorption cycle, rendering it more susceptible to changes in heat input.

In summary MIL-101(Cr) offers several advantages over its counterpart, activated carbon. Nonetheless, it is essential to acknowledge some limitations of MOFs that have been considered previously. Therefore, further research and development are required to enhance the feasibility of MOF utilization in the future.

8.8 Conclusion

In conclusion, this paper presents a comprehensive literature review on MOFs and their application in heat adsorption transformation. Particularly, MIL-101(Cr) has exhibited exceptional ammonia sorption stability, high adsorption performance and minimal hysteresis. Therefore, it was chosen to be investigated for its efficacy in a 10 kW cooling capacity system, and the system's coefficient of performance was compared against activated carbon. MIL-101(Cr) exhibited superior performance compared to activated carbon, as the system required a lower mass of adsorbent to produce 10 kW of power and consumed less heat input while demonstrating a higher coefficient of performance. The findings of this study have contributed to a deeper understanding of MOFs as potential candidates for efficient and effective heat adsorption in refrigeration applications.

This study has shown that the COP of MIL-101(Cr) is about 1.5 times more than that of activated carbon, highlighting the potential of MOFs in the field of refrigeration. Nevertheless, it is important to note that MOFs are still in the early stages of development and require extensive research before they can be used in residential refrigeration applications. This is because the production and manipulation of MOFs is challenging, and their current limited supply makes their commercial viability uncertain. More research and technological breakthroughs are necessary to overcome these challenges and to prove that they are a more efficient alternative to current refrigeration technologies.

References

1. Heating and cooling, *MIT Climate Portal*. [Online]. Available: https://climate.mit.edu/explainers/heating-and-cooling. Accessed: 13 Mar 2023
2. K. Wang, E.A. Vineyard, Adsorption refrigeration system. ASHRAE J. 01 Jan 2011. [Online]. Available: https://www.osti.gov/biblio/1024270#:~:text=Adsorption%20refrigeration%20is%20an%20environmentally,high%20temperature%20and%20salt%20crystallization. Accessed: 08 Mar 2023
3. C.Y. Yan, 6.2 Refrigerator and Heat Pump, Introduction to Engineering Thermodynamics. [Online]. Available: https://pressbooks.bccampus.ca/thermo1/chapter/6-2-refrigerator-and-heat-pump/. Accessed: 08 Mar 2023
4. R. Gopal, What is Absorption?—Definition from Corrosionpedia, Corrosionpedia. [Online]. Available: https://www.corrosionpedia.com/definition/26/absorption. Accessed: 08 Mar 2023
5. Adsorption, Adsorption An Overview| ScienceDirect Topics. [Online]. Available: https://www.sciencedirect.com/topics/earth-and-planetary-sciences/adsorption. Accessed: 08 Mar 2023
6. X.-J. Kong, J.-R. Li, An overview of metal–organic frameworks for green chemical engineering. Engineering **7**(8), 1115–1139 (2021)
7. E. Hastürk, S.-J. Ernst, C. Janiak, Recent advances in adsorption heat transformation focusing on the development of adsorbent materials. Curr. Opin. Chem. Eng. **24**, 26–36 (2019)
8. S. Wang, X. Xia, S. Li, Cooling performance of metal organic framework-water pairs in cascaded adsorption chillers. Appl. Therm. Eng. **189**, 116707 (2021)
9. M. Berger, What is a MOF (metal organic framework)? Nanotechnology 02 Jul 2021. [Online]. Available: https://www.nanowerk.com/mof-metal-organic-framework.php. Accessed: 08 Mar 2023
10. A. Rocchetti, M. Lippi, L. Socci, P. Gullo, V. Khorshidi, L. Talluri, Metal-organic framework adsorbent materials in HVAC systems: general survey and theoretical assessment. Energies **15**(23), 8908 (2022)
11. S. Vasta, V. Brancato, D. La Rosa, V. Palomba, G. Restuccia, A. Sapienza, A. Frazzica, Adsorption heat storage: state-of-the-art and future perspectives. Nanomaterials **8**(7), 522 (2018)
12. A. Rezk, R. Al-Dadah, S. Mahmoud, A. Elsayed, Characterisation of metal organic frameworks for adsorption cooling. Int. J. Heat Mass Transf. **55**(25–26), 7366–7374 (2012)

13. N. Riaz, M. Sultan, S. Noor, U. Sajjad, M. Farooq, M.U. Khan, S. Hanif, F. Riaz, Recent developments in adsorption heat pumps for heating applications. Adv. Mech. Eng. **14**(4), 168781322210894 (2022)
14. M. Moayed Mohseni, M. Jouyandeh, S. Mohammad Sajadi, A. Hejna, S. Habibzadeh, A. Mohaddespour, N. Rabiee, H. Daneshgar, O. Akhavan, M. Asadnia, M. Rabiee, S. Ramakrishna, R. Luque, M. Reza Saeb, Metal- organic frameworks (MOF) based heat transfer: a comprehensive review. Chem. Eng. J. **449**, 137700 (2022)
15. P. Henshaw, J. Aman, D.S. Ting, Solar sorption cooling for residential air-conditioning applications. Int. J. Renew. Energy Technol. **9**(1/2), 136 (2018)
16. G. An, X. Xia, S. Wu, Z. Liu, L. Wang, S. Li, Metal–organic frameworks for ammonia-based thermal energy storage. Small **17**(44), 2102689 (2021)
17. Metal organic frameworks, *Orange* (Central West Publishing, Australia, NSW, 2019)
18. X. Yang, A. Arami-Niya, J. Lyu, X. Guo, Net, excess, and absolute adsorption of N_2, CH_4, and CO_2 on metal–organic frameworks of zif-8, MIL-101(CR), and uio-66 at 282–361 K and up to 12 MPA. J. Chem. Eng. Data **66**(1), 404–414 (2020). https://doi.org/10.1021/acs.jced.0c00738
19. MOFs vs. Other Porous Materials for Carbon Capture, novoMOF (2022). Available at: https://blog.novomof.com/mofs-vs.-other-porous-materials-for-carbon-capture Accessed: 28 April 2023

Chapter 9
Thermal Analysis of the Nacelle of a Small Horizontal Axis Wind Turbine Using a CFD Model in ANSYS-FLUENT

Bhaswati Sen and Nabanita Datta

Abstract Horizontal Axis Wind Turbines (HAWT-s) are the most commonly used wind turbines all over the world in the sector of renewable energy. Apart from the blades, the nacelle is an important part of HAWT-s, which houses the electromechanical equipment. However, the nacelle being a closed chamber, mostly sealed from the outside, is prone to temperature increase due to the heat generated by the drivetrain, brake and, most significantly, generator. Commercial, high-capacity HAWT-s employ "forced cooling" systems to tackle this issue and published literature indicates the application of computational fluid dynamics (CFD) software packages in studying the effectiveness of the cooling techniques for the nacelles of such turbines. For small-capacity, low-cost HAWT-s meant for community level power supply, the cooling of the nacelle may be effected by a zero-cost natural ventilation system in the form of a pair of air vents on the windward and leeward sides, that is, the front and rear faces of the nacelle respectively, through which a draft of air is allowed to pass for cooling the nacelle components. However, CFD analysis for such systems have so far gone unreported, and this study aims at filling up the gap by making use of the software package ANSYS FLUENT (student version) for predicting the air flow pattern and temperature variations within the nacelle and evaluating the effectiveness of the air vents in keeping the temperatures sufficiently in check. It must be emphasized that the CFD studies reported in literature for larger wind turbines have mostly been demonstrated for two-dimensional computation domains considering only the space inside the nacelle body. The present work goes a step further considering a larger three-dimensional computation domain encompassing the space bounded by the nacelle, and connecting the two through the pair of air vents. This helps in simulating the free stream of air outside the nacelle and also helps in modelling the draft of air entering the nacelle through its front and leaving from behind. The simulation

B. Sen (✉)
Department of Civil Engineering, Indian Institute of Engineering Science and Technology (IIEST), Shipbur, West Bengal, India
e-mail: sen.bhaswati2020ce@gmail.com

N. Datta
Department of Ocean Engineering and Naval Architecture, Indian Institute of Technology (IIT), Kharagpur, West Bengal, India

D. S-K. Ting and A. Vasel-Be-Hagh (eds.), *Engineering to Adapt*, Springer Proceedings in Energy, https://doi.org/10.1007/978-3-031-47237-4_9

results indicate that the temperature does not reach very high values in the region from 0.2 to 0.4 m above the bottom of the nacelle, considering the nacelle to have a height of 0.41 m. This region, therefore, is suitable for the placement the electronic equipment necessary to control the wind turbine movements and operation. This work also reveals that the natural cooling attained by the draft of outside air directed into the nacelle is able to maintain the temperature in the nacelle well within the acceptable limits. The provision of circulation vents of suitable diameter on the front and rear faces of the nacelle, therefore, could provide a low cost solution to the cooling mechanism of the nacelle in small HAWTs.

Keywords Computational Fluid Dynamics · Wind turbine nacelle · Thermal analysis · Natural ventilation cooling

9.1 Introduction

There is a growing concern today about the impact of utilizing fossil fuel for power generation, propulsion and other human activities on the earth's climate. Indeed, many countries have shifted their power production technology towards greener sources in recent times. Even countries like Canada, endowed with an overwhelming source of hydropower contributing to 60% of its total power generation, has invested in non-hydro renewable sources, which presently stands at around 8% [1]. Countries like India, which had a traditional reliance on coal for grid power generation, has now increased its power production from non-hydro renewable sources to 30% [2]. Cooling of the nacelle box, which houses the heat generating electromechanical components of the wind turbine, is an important aspect of wind turbine design. Commercial high capacity turbines use forced cooling technology and several studies are available that make use of Computational Fluid Dynamics (CFD) methods to analyse its efficiency [5, 6, 15–22]. However, none of these studies consider an interaction between the air domain surrounding a nacelle and the air within, as has been demonstrated in this paper. Moreover, many of the references cited consider a CFD domain only in two-dimension (2D), whereas a complete three-dimensional (3D) domain is modelled for the study reported herein. Finally, this paper reports for the first time, a CFD study on natural ventilated cooling for wind turbine nacelles with the help of air vents placed strategically on the front and rear faces of the nacelle box.

The horizontal-axis wind turbine was tested in the Institute campus at Kharagpur, because of its proximity to design and fabrication facilities. However, since the location is around 100 km away from the coastline, it does not enjoy high wind speeds at the ground level. Hence, the turbine is being tested on the roof of a 10-storey building within the campus, in order to harness the slightly higher wind speeds available at this level.

The design wind speed of the turbine, however, is higher as may be expected along the coastline of the country.

Figure 9.1 shows a schematic diagram of the horizontal-axis wind turbine considered in this paper. The main component that is the focus of this paper is the nacelle, or the enclosure that houses all the major components of a wind-turbine, except the blades. The purpose of the nacelle is to safely protect different components of a wind turbine such as the gear-box, coupling, generation unit, brakes, and the electronic control system that monitors the power input/output as well as the controls the direction of the blades according to the direction of the prevailing wind. It also controls the motor when it is required to provide the initial torque to start the turbine and again helps in deciding when to apply the brakes when the wind speed crosses a threshold. The electronic components are sensitive to temperatures and need to be protected from the heat generated within the nacelle by the gearbox, brake and induction motor that also doubles up as the generator. It may be noted that the wind speed and direction are obtained using a wind vane and a wind anemometer located outside the turbine but the signals are led through cables to the electronic components within the nacelle. For controlling the yaw of the turbine, the prevailing wind direction is also sensed by the control system and the slewing motor is driven for aligning the turbine axis in the desired direction. The direction is altered, if need be, with the active control of the system as programmed in the electronic equipment placed within the nacelle.

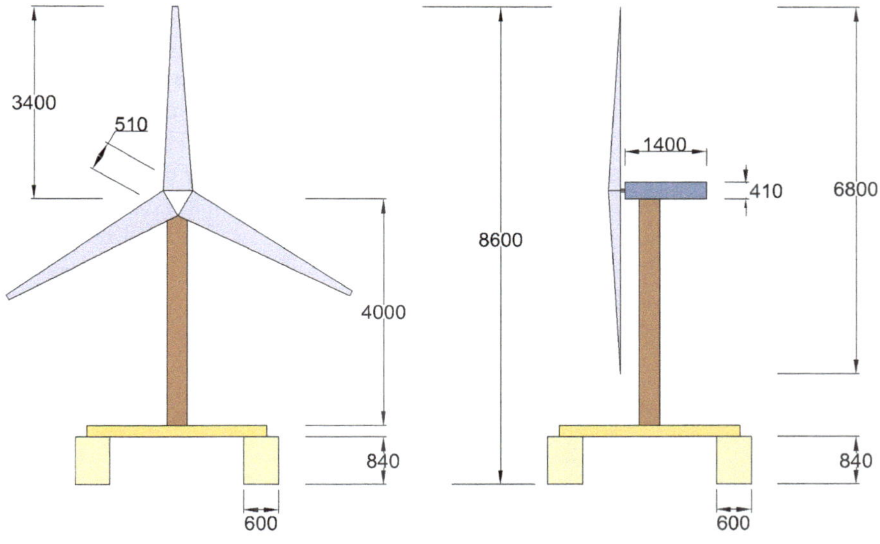

Fig. 9.1 Salient dimensions of the horizontal-axis wind turbine considered in this study

9.2 Cooling of the Nacelle: Techniques, Solutions and Motivation for the Study

Considering the risk of damage to the temperature sensitive electronic and other equipment placed inside the nacelle due to the heat generated by the different moving components of the drivetrain and generator housed within the nacelle, all wind turbines invariably use some form of a cooling system to regulate the temperature rise and maintain it within acceptable limits. Commercial turbines, with higher rated power, use fans and other forced cooling systems to achieve temperatures tolerable to the electronic control system [3, 4, 25]. However, the present project aims at developing a low-cost wind turbine for local communities and thus cannot afford to expend additional energy for the cooling of the nacelle interior. The enclosed volume of the nacelle is small, thus limiting the space available for heat dissipation within the nacelle body generated by the electromechanical components. Hence, practitioners of low-cost small wind turbines have suggested providing air vents on the front and rear faces of the nacelle body, thus permitting a flow of air into the nacelle and aiding the cooling of the components naturally. The optimum location and size of vents may be assessed by the use of computer computational fluid dynamics (CFD) studies to ensure the effectiveness of the air flow in dissipating the generated heat. Studies reported so far, carried out mostly for commercial wind turbines with forced cooling, have used two dimensional computation domains in carrying out their analyses and that too considering only the nacelle interior. The present study is more comprehensive, taking into account a three dimensional model domain much larger than the nacelle, thus encompassing the nacelle geometry. It is thus able to model the interaction of the flow taking place in the free stream air and the portion of which that enters and leaves the nacelle through the air vents. This study demonstrates for the first time, using CFD simulations, the effectiveness of air vents on the dissipation of generated heat inside a nacelle that is cooled naturally with a draft of air. The following section reviews briefly some aspects of CFD modelling in understanding the flow and heat dissipation within a nacelle, and makes a case for the naturally ventilated cooling method described in the present work.

9.3 Computer Modelling and Analysis of Flow and Heat Transfer in the Nacelle Interior: A Review

Although CFD simulation of the nacelle interior of a small wind turbine has not been reported yet in mainstream journals, research studies on the use of CFD tools to study air flow on the outside of the nacelle and air flow together with heat dissipation within the body of the nacelle of wind turbines have appeared since the turn of the new millennium. Masson et al. [5] probably were one of the few early researchers to apply a CFD code to study the dynamic behavior of rotor aerodynamic characteristics. They employed the incompressible Reynolds Averaged Navier–Stokes equations for simulating the flow field on the rotor of a horizontal axis wind turbine and estimated its drag and lift. The control-volume finite-element method (CVFEM), developed for an axisymmetric polar case by Masson et al. [6] and for a general three-dimensional Cartesian domain by Saabas and Baliga [7], were employed in this study. Although this work investigated only the aerodynamics, other publications have appeared over the course of time that investigated flow movement and heat exchange both inside and outside of a nacelle. Most of these studies had followed the methodology of flow simulation around bluff bodies [8–13].

Using the above-mentioned numerical model CVFEM, Smaili and Masson [14] investigated the impact of the moving rotor blades of a wind turbine on the air flow characteristics over the nacelle. Shortly thereafter, the same group of researchers investigated the thermal behavior of a wind turbine nacelle assumed to be operating under Nordic conditions. Published by Smaili et al. [15], the researchers used the references mentioned in the previous paragraph as the basis for analyzing the thermal behavior of a wind turbine nacelle assumed to be operating under Nordic conditions. They simulated both the airflow outside as well as inside of the nacelle by solving the Reynolds-averaged Navier–Stokes equation using the unstructured grid control CVFEM. The standard k-ε model was used for the description of turbulent flow. The simulations were carried out for a commercial 750 kW wind turbine. The simulation results suggest that to maintain an acceptable temperature level within the nacelle, the air flow rate needs to be adjusted as a function of wind velocity and external air temperature. They also recommended the use of a fan for air ventilation cooling. Tran et al. [16] extended the original CVFEM formulation with a second-order upwind scheme for convection terms and implemented in the context of CVFEM methodology.

Since around 2010, several CFD studies have come up that deal with exclusively on the flow and heat transfer within the enclosed space of a nacelle, with an aim to simulate the convection and dissipation of the heat generated by the heat producing sources such as the gearbox, brake and generator. For example, Smaili et al. [17] used the CVFEM technique to simulate the air flow and heat movement within a nacelle of a wind turbine that is expected to withstand the extreme heat of Algerian Saharan desert. They used the Reynolds-averaged Navier–Stokes equation along with the standard k-ε model for turbulence closure. The researchers assumed the nacelle to be air-tight with the heat being generated inside being dissipated within the nacelle body itself. Only the generator, which is the major source of heat, was considered in the study and heat source was idealized as a hot plate at a fixed temperature.

Tekin et al. [18] used the ANSYS-FLUENT commercial CFD code to study the thermal behavior of the nacelle of a large wind turbine (2.5 MW) under cold weather. In this case, a closed system application was considered with no flow circulation within the nacelle. A two-dimensional domain was considered and the Spalart–Allmaras' k-ε turbulence closure model was used in the CFD code. The authors used the CFD simulations to understand the surface temperature distribution over the components in the nacelle under different air conditioning capacities. Mahdi and

Smaili [19] also used ANSYS-FLUENT to analyze the heat dissipation by the generator inside the nacelle of a medium sized (850 kW) wind turbine by considering a two-dimensional domain and using structured rectangular grid. The Reynolds Averaged Navier–Stokes (RANS) equations were solved by the model using the shear-stress transport (SST) k-ω turbulence closure model. Bhamare and Gadhe [20] used the OpenFOAM public domain CFD software package to model the nacelle of a large capacity (1 MW–6 MW) wind turbine using a three-dimensional structured grid. The CFD model uses the k-ε turbulence closure model and in common with similar studies, a dynamically achieved steady state condition was assumed. The authors have reported the results of an investigation into the temperature distribution and velocity vectors within the nacelle of the turbine under three configurations of forced air-driven nacelle cooling systems. Further, they studied the effect of increasing the power of the cooling fans on the temperature distributions. Mahdi et al. [21] used the ANSYS-FLUENT CFD code to model in two and three-dimensions the impact of turbulent natural convection heat transfer on the thermal behavior of the nacelle of a wind turbine under hot temperature settings. Zhang et al. [22] discussed the numerical simulation of turbulent natural convection heat transfer in a high power wind turbine of the order of 10 MW. Using both two and three-dimensional models of the nacelle domain, the RANS equations were solved along with the SST k-ω turbulence closure model. Alkaabi [23], has carried out in detail as a part of a master thesis, the CFD analysis on the cooling of a HAWT Nacelle. The CFD open source code ANSYS-FLUENT [24] was used to model the nacelle geometry of a 2.5 MW rated wind turbine in three-dimension and three configurations for cooling the nacelle interior are tested. One objective of the thesis work was to compare the efficacy of a forced air-cooling system versus a liquid cooling system and identify the most effective strategy for cooling the nacelle.

9.4 Computer Simulation of Natural Ventilation Cooling in Small HAWT Nacelle

The present work discusses the effect of natural ventilation taking place through two sets of vents on the nacelle walls on dissipating the temperature inside the nacelle of a small low-capacity horizontal axis wind turbine using the commercial CFD code ANSYS-FLUENT, student version [24]. The vents allow a natural draft of air to flow through the vents, thus helping to dissipate a part of the heat generated by the electromechanical components of the wind turbine that is housed within the nacelle. The study is undertaken for a student-designed captive off-grid wind generated power generation project for meeting the minimum energy demands of a typical small community (details given in Table 9.1). Figure 9.2 shows a schematic diagram of the nacelle box where, the main source of heat generation is the induction motor-cum-generator, position of the electronic control equipment, and the front and rear vent holes to permit air flow through the nacelle. The other components, such as the gear box, drive train, coupling mechanism, generator, etc. are not indicated for simplicity.

Since the final designed product is proposed to be kept affordable to the weaker sections of the society, a natural ventilation cooling system has been proposed that would not require expensive fans or air-coolers to lower the temperature of the air enclosed within the nacelle of the wind turbine. However, an overall assessment needs to be done to ensure that the proposed natural air draft would be capable of achieving a temperature low enough to prevent damage to the sensitive electronic components enclosed within the nacelle from the heat produced by the generator and other components. It may be worth mentioning that the induction motor is close to the rear of the rotor shaft and near the floor of the nacelle. The position of the electronic equipment is, therefore, kept deliberately placed on the ceiling of the nacelle and more towards the front end, away from the main source of heat as much as possible. Further, the wind is expected to blow in from the front vents and leave from the rear, thus the heat generated by the induction motor should be convected by the flowing air, away from the electronic equipment cabinet. Since the CFD software was used to understand the flow and heat transfer within the nacelle due to natural air circulation, the following governing equations, The Reynolds Averaged Navier–Stokes (RANS) equations and the conservation of energy equation (in the steady state) which are solved by the CFD code is given below, written in Cartesian tensor notation:

Table 9.1 Salient features of a horizontal-axis experimental small wind turbine being designed and fabricated by students at Indian Institute of Technology Kharagpur, India

Rotor diameter	6.8 m
Hub height (above base plate)	4.0 m
Design (or rated) wind speed	4.2 ms^{-1}
Design power output	250 W
Cut-in speed	2.1 ms^{-1}
Cut-out speed	9.8 ms^{-1}

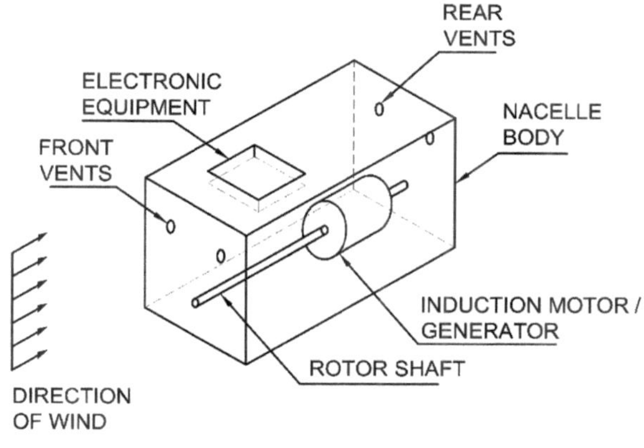

Fig. 9.2 Schematic view of the nacelle considered in the study, showing the placement of components

$$\frac{\partial(\rho u_i)}{\partial x_i} = 0 \tag{9.1}$$

$$\rho u_j \frac{\partial u_i}{\partial x_j} = -\frac{\partial p}{\partial x_i} + \frac{\partial}{\partial x_j}\left(\mu_e\left(\frac{\partial u_i}{\partial x_j} + \frac{\partial u_j}{\partial x_i}\right)\right) \tag{9.2}$$

$$\rho u_j \frac{\partial\left(c_p T\right)}{\partial x_j} = \frac{\partial}{\partial x_j}\left(k_e \frac{\partial T}{\partial x_j}\right) + \mu_e\left(\frac{\partial u_i}{\partial x_j} + \frac{\partial u_j}{\partial x_i}\right)\frac{\partial u_i}{\partial x_j} \tag{9.3}$$

where, u_i is the ith flow velocity component, T is the air temperature, p is the air pressure, ρ and c_p are density and heat capacity of the air, respectively. To take into consideration the temperature dependence of the air density, the ideal gas law has been used as:

$$\frac{p}{\rho} = R_{air} T \tag{9.4}$$

with R_{air} is the specific air constant.

In the above equations, $\mu_e = (\mu + \mu_t)$ and $k_e = (k + k_t)$ refer respectively to effective viscosity and effective thermal conductivity; in which, μ is the molecular viscosity and μt is turbulent viscosity, and k is the molecular thermal conductivity and k_t is turbulent thermal conductivity. The turbulent properties (μ_t and k_t) can be estimated from an appropriate turbulent model.

9.5 Numerical Solution to the Distribution of Velocity and Temperature Within the Nacelle

The CFD code ANSYS-FLUENT, employed in this study, uses the finite volume technique to solve the set of Eqs. (9.1) to (9.3), mentioned above, for arriving at a solution. Although the equations are time dependent, the boundary conditions are given as time invariants which allow the converged solutions to attain a steady state that is in a dynamic equilibrium with the surroundings.

The present study explores the distribution of velocity and temperature within the nacelle, due to the natural ventilation effected by the air entering the nacelle from outside through the front vents, and then leaving through the rear vents. This is a three dimensional phenomenon, which is not possible to study using a two-dimensional domain, as reported in several earlier studies [15, 17, 19, 20]. Hence, a larger domain of air, in three-dimensions, is considered within which the cuboidal domain representing the nacelle is placed (Fig. 9.3a). Two circular vents, diameter 30 mm diameters each, are modelled on the front vertical face of the nacelle. Similar vents are modelled on the rear face of the nacelle as well. Since the nacelle is aligned perpendicular to the plane of the turbine blades, the air surrounding the nacelle is assumed to blow face on towards the nacelle, having an average free-stream velocity of 6 ms^{-1}. An ambient temperature of 30 °C is assumed to prevail in the air surrounding the nacelle. The location of the air vents is decided from practical considerations by accounting for all possible alignments of the blade and considering the location which remains uncovered by the various combinations of blade movement. The nacelle box is assumed to be lying with its base plate aligned with the bottom horizontal surface of the larger domain of air. The larger air domain enclosing the smaller nacelle body and the air within the nacelle are both decomposed into three-dimensional elements, for the sake of numerical calculations to be performed by the software package. Note that the two air domains are connected through passages defined by the two sets of vents, one each on the front and the rear faces of the nacelle. Unstructured meshing, using three dimensional tetrahedral elements of varying size, is adopted since detailed analysis is required at specific locations. Moreover, the air vent passages are relatively very small compared to the larger domain, and require smaller mesh size for proper resolution of the computed variables. The larger bounding air domain, however, is decomposed with a coarser mesh since it serves to create the free-stream of air flowing from the front and around the nacelle. Figure 9.3b shows the two-dimensional imprint of the three-dimensional mesh on the surface of the nacelle. The nacelle body is made of aluminum and defined accordingly in the computation software. The electromechanical components of the wind turbine; such the rotor shaft, gear box, coupling, generator, etc. are considered to be made of steel. The primary source of heat, the induction motor-generator, is estimated to emanate 4724 Wm^{-3} corresponding to the loss of energy expected of the machine running under the conditions, and is included as heat source in the computational model.

The boundary conditions for running the computational model in the direction of wind flow are specified thus: (a) specified free-stream wind velocity at the inlet surface; (b) atmospheric pressure condition in the air-outlet surface, located opposite to the inlet surface. In addition, there are four lateral no-flow boundary surfaces, on which atmospheric pressure condition is assumed to prevail. The meshed domain inside the nacelle is connected with the outside through the front and rear vents.

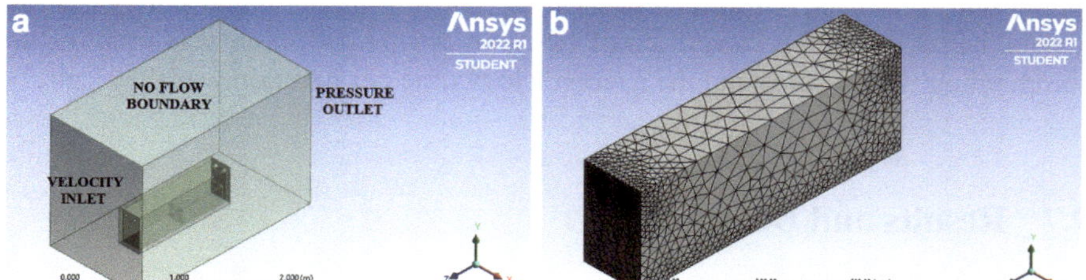

Fig. 9.3 **a** Computational domain enclosing the nacelle cuboid with boundary conditions. Air flow assumed to be in the negative z-direction. **b** Close-up view of the domain encompassing the nacelle, showing the computational mesh footprint on the nacelle surface

While carrying out the computer simulations using ANSYS-FLUENT, the appropriate computational models for simulating air flow and heat transfer are chosen. The boundary conditions are also appropriately applied, such as constant uniform velocity on the inlet face and atmospheric pressure at the outlet face. The other four faces are specified as no-flow boundaries. The unsteady state simulations are run till the residuals for the errors reduce to within an acceptable threshold. At this stage, a dynamic equilibrium is reached, signifying the steady state condition.

9.6 Grid Optimization-Mesh Convergence Study

It is accepted that finer the mesh size, that is, the largest edge length of the elements used in discretizing the domain, yields more accurate results. However, finer mesh sizes also call for a greater computational time. This issue becomes even more important when the simulations are run on personal computers, which may take immensely long time if mesh size is greatly refined. The simulations, therefore, are required to be carried out for an optimum grid size, which means that it is neither too coarse to predict unacceptable results, and at the same time, is not too fine which causes an intense pressure on the computer resources. In all there were 236,304 tetrahedral elements.

Accordingly, three mesh sizes are tested, which are 0.144 m, 0.100 m and 0.05 m respectively, with the mesh getting finer and number of cells increasing with very reducing average mesh size. The temperature results at a specific location within the nacelle are then captured for each mesh and the results plotted in Fig. 9.4, which shows that the mesh size of 0.10 m was the optimum since a finer mesh produced practically the same result as with this mesh size, whereas a coarser mesh produces a different temperature value.

It may be also mentioned that the numerical computations in ANSYS-FLUENT are carried out iteratively, at each time step. In the present study, a maximum iteration value of 1000 has been set. A converged solution is accepted if the error residuals monotonically reduce with the iterations, as shown in Fig. 9.5, and finally reaches a value below a specified threshold of 10^{-5}. An *intel i5* processor was used for the computation with 4 cores and the average clock time per iteration was 0.391 ms.

9.7 Results and Discussion

The results of the three dimensional domain are shown in this section through two vertical planes passing through the computational domain and cutting through: (a) the centerline of the nacelle (Plane 1); and (b) front and rear vent holes (Plane 2). The two planes are shown in Fig. 9.6.

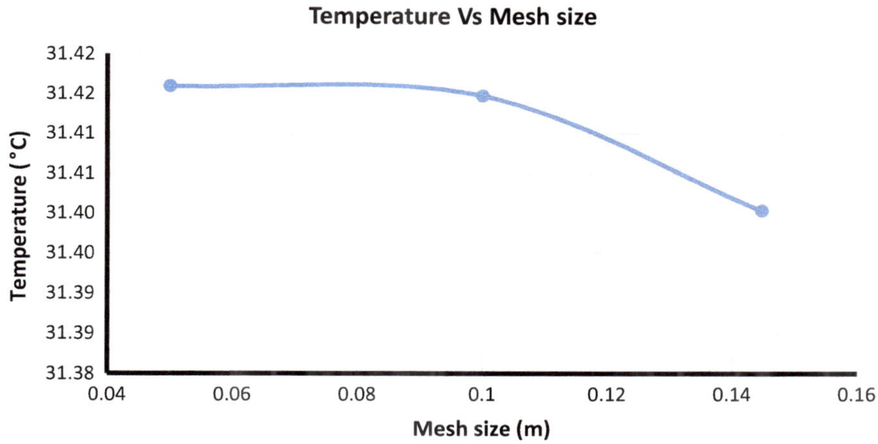

Fig. 9.4 Temperature at the center of the ceiling inside the nacelle, plotted against mesh grid size

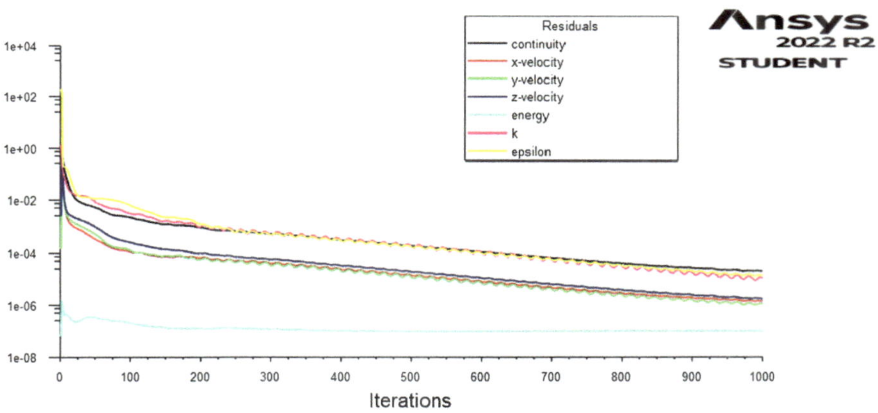

Fig. 9.5 A typical graph showing the reduction of error residuals with iterations

The results obtained are presented in three categories, as indicated below:

1. Contours of velocity magnitude,
2. Vectors showing velocity direction, and
3. Contours of temperature magnitude.

Figure 9.7a and b show the velocity magnitude contours through the cross sectional planes 1 and 2, indicated in Fig. 9.6. From Fig. 9.7b, it is evident that a stream of air enters through the front vents of the nacelle, gets dispersed within the nacelle body, and leaves through the rear vents as a stream. Although Fig. 9.7a shows the velocity vectors on the centerline plane of the nacelle, and not through one of the vents, a circulation of air is apparent from the velocity contours due to the three dimensional

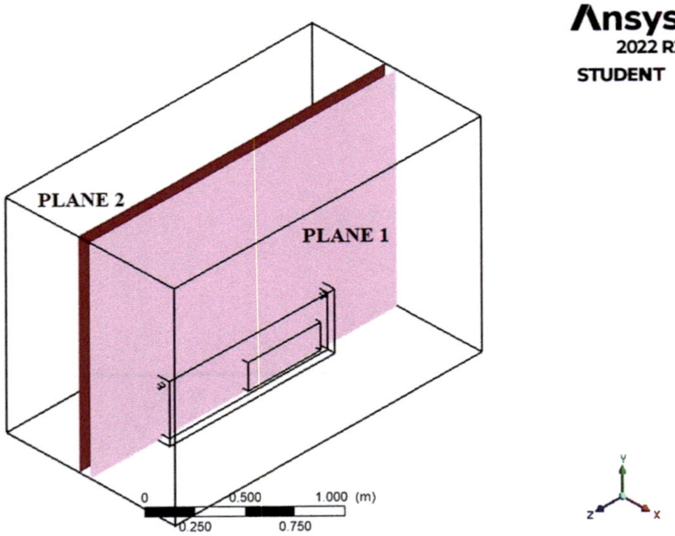

Fig. 9.6 Sectional planes for visualization through the nacelle; Plane 1 passes through the centerline of the nacelle, while Plane 2 passes through the vents

effect. Since the nacelle is not aerodynamically streamlined, the outside top leading edge of the nacelle experiences a stagnation point with the jet of air flowing over it. It may be noted that the outline of the induction motor, which acts as the source of heat generation, is indicated by a solid rectangle in Fig. 9.7a and b.

The velocity vector through Plane 2 is showed in Figs. 9.8 and 9.9 shows the temperature contour through Plane 2. A circulation of air inside the nacelle is seen to occur, driven by the streams of air entering the nacelle through the front vents. This is extremely useful for dissipating the heat generated by the induction motor, and for also carrying some of the heat away through the rear vent by way of convection. This is confirmed from the temperature contours as shown in Fig. 9.9, which indicates that the maximum temperature rise of the air inside the nacelle compartment is 39.14

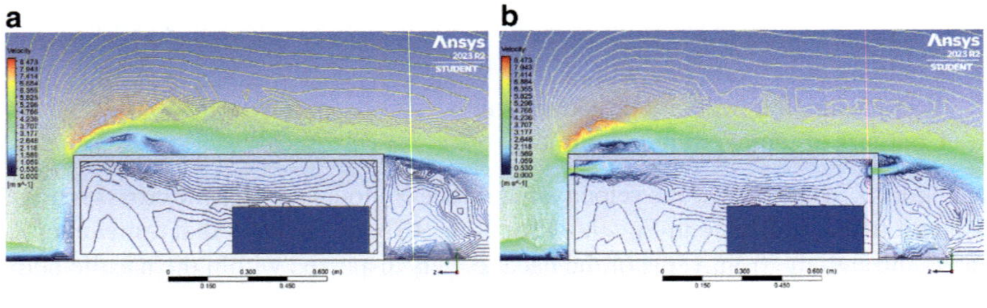

Fig. 9.7 **a** Velocity contours through sectional Plane 1. **b** Velocity contours through sectional Plane 2

°C, just around the periphery of the induction motor. In the remaining space within the nacelle, including at the location of the electronic equipment, the temperature is only slightly above the ambient temperature of 30 °C.

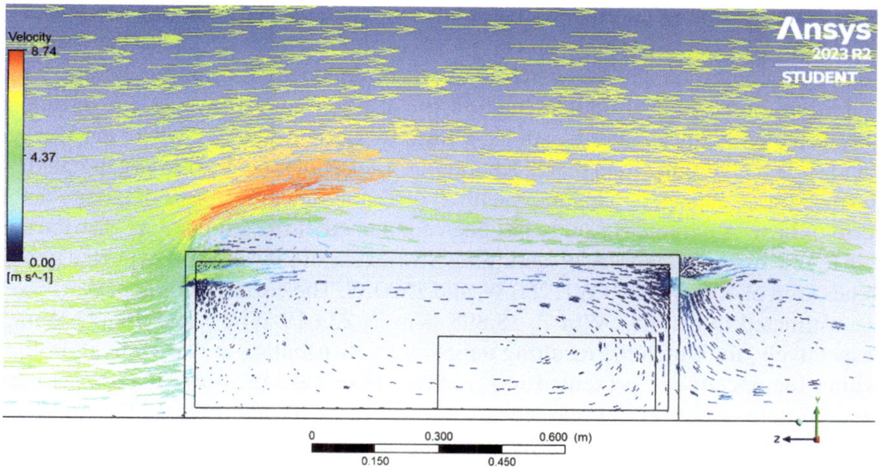

Fig. 9.8 Velocity vectors through sectional Plane 2

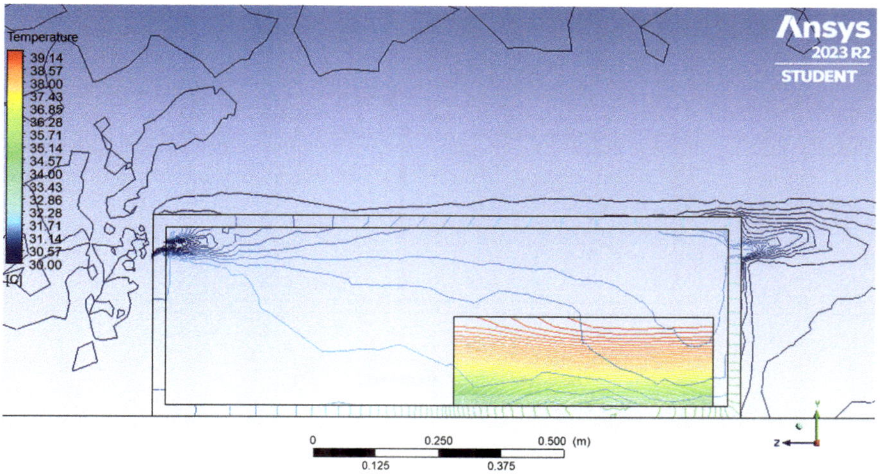

Fig. 9.9 Temperature contours through sectional Plane 2

9.8 Parametric Study

The computer simulations shown above for the wind turbine nacelle air flow and temperature distribution have been carried out for a free stream air velocity of 6 $ms^{-1}$and assuming that the induction motor-generator is dissipating 4724 Wm^{-3} of power. This corresponds to an energy loss from the machine under the assumed power generation of 250 W, which is expected under the prevailing average wind conditions at the project implementation site at Kharagpur. However, the turbine is also designed to run at other locations, especially along the coasts, where the average wind speeds are higher than that at Kharagpur. Consequently, the power generated by the machine is expected to increase, with a corresponding greater loss of energy in the form of heat. A parametric study has been conducted of the temperature distribution due to the higher heat losses by assuming a 30% loss of power, from expected power generation of 500 W, 1000 W, 1250 W, and 1500 W. The corresponding power losses are estimated to be 9448 Wm^{-3}, 18,895 Wm^{-3}, 23,619 Wm^{-3} and 28,343 Wm^{-3} respectively, for which the resulting temperature distribution at two vertical profiles within the nacelle are presented in Fig. 9.10. Figure 9.11a shows the temperature variation along the profile marked as A-A in Fig. 9.10, and Fig. 9.11b shows the same temperature profile along the plane marked as B-B in Fig. 9.10.

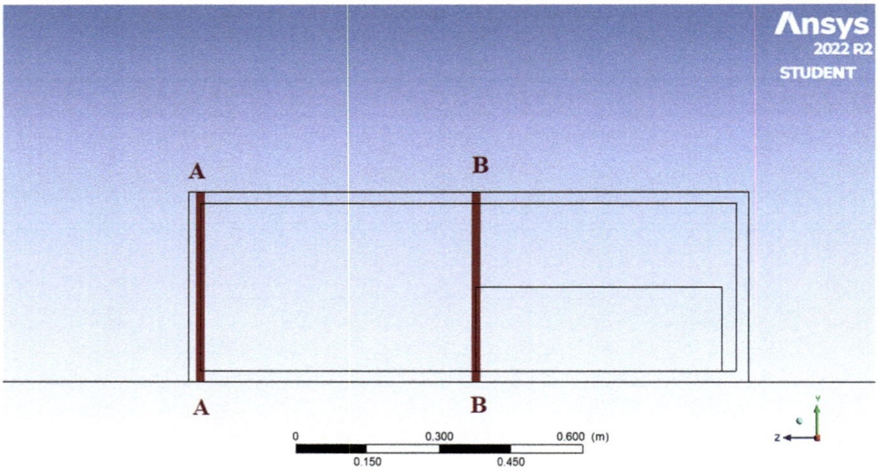

Fig. 9.10 Location of the two vertical profile planes for representation of temperatures

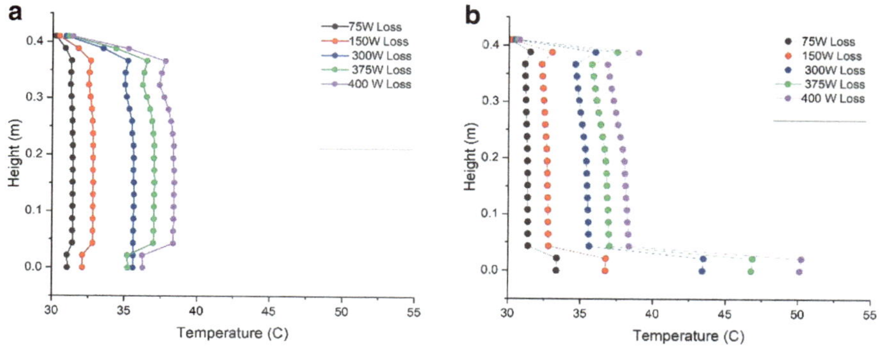

Fig. 9.11 **a** Temperature through profile Plane A-A. **b** Temperature through profile Plane B-B

9.9 Concluding Remarks

The present study reported in this paper discusses the use of a CFD simulation code, ANSYS-FLUENT, for determining the distribution of velocity and temperature profile inside the nacelle of a small horizontal axis wind turbine (HAWT). A small HAWT is said to be one which has a swept area of less than 200 m^2 and a rated power of up to 50 kW. Since the generator of a wind turbine acts as a heat source while in operation, it is essential to keep the air temperature within the nacelle body within an acceptable limit in order to avoid damage to other sensitive equipment located within the nacelle. Air vents provided in the nacelle permitting outside air to enter and moderate the temperature are checked with the CFD code to see whether the size and location of the vents are sufficient to achieve this goal. Results reveal that the temperature rises rapidly and at a greater magnitude in the region in the immediate surroundings of the induction motor as expected. It also shows that the temperature reaches almost a constant low for all the cases in the region from 0.2 to 0.4 m. By comparing the results obtained from different heat loss scenarios it can be suggested that, from protection point of view, the region between 0.2 and 0.4 m is suitable for the placement the electronic equipment. The analyses conducted in this study also reveal that the natural cooling attained by the draft of outside air directed into the nacelle does indeed maintain the temperature in the nacelle well within the acceptable limits. Thus for small wind turbines adopting circulation vents of suitable diameter based on the nacelle size could be a novel approach to the cooling mechanism of the nacelle.

9.10 Scope of Further Work

Although the natural ventilation assisted cooling is seen to work well for a low-power HAWT, for higher wind speeds and correspondingly higher power generation and heat loss, it may become necessary to adopt forced cooling systems, as seen in commercial turbines [25]. As an extension of this work, therefore, a cooling system may be designed and tested which, should be economical while, being effective in handling the higher heat dissipation within the nacelle for higher power outputs. The effect of radiation could also be a further addition to the current direction of work.

Acknowledgements The first author gratefully acknowledges the support of MITACS Globalink Summer Research Internship for an internship opportunity at the University of Windsor, Canada, and particularly for the participation in this conference. She also hereby expresses gratitude to her fellow team-mates of the Chakra Sindhu Wind Turbine Project at IIT Kharagpur who generously helped in gathering relevant data required for this study.

References

1. Natural Resources Canada, Energy Fact Book 2022–2023 (2022)
2. Ministry of Power—Government of India, Power Sector at a Glance, online resource https://powermin.gov.in/en/content/power-sector-glance-all-india (2023)
3. S. Jiana, M. Xiaoqianb, C. Shuyingc, G. Huijing, Review of the cooling technology for high-power wind turbines, in *5th International Conference on Advanced Design and Manufacturing Engineering* (2015)
4. O. Nematollahi, K. Chung Kim, Wind turbine cooling: the state-of-the-art review, in *Proceedings of Academics World International Conference*, Tehran, Iran, 12th–13th (2017)
5. C. Masson, A. Smaili, C. Leclerc, Aerodynamic analysis of HAWTs operating in unsteady conditions. Wind Energy **4**, 1–22 (2001)
6. C. Masson, H.J. Saabas, B.R. Baliga, Co-located equal-order control-volume finite element method for two-dimensional axisymmetric incompressible fluid flow. Int. J. Numer. Methods Fluids **18**(1), 1–26 (1994)
7. H.J. Saabas, B.R. Baliga, Co-located equal-order control-volume finite-element method for multidimensional, incompressible. Fluid Flow-Part I: Formul. Numer. Heat Transf. **26**(4), 381–407 (1994)
8. F.-S. Lien, E. Yee, Numerical modelling of the turbulent flow developing within and over a 3-d building array, part I: a high-resolution Reynolds-averaged NavierStokes approach. Boundary-Layer Meteorol. **112**(3), 427–466 (2004)
9. X. Zhang, A. Senior, A. Ruhrmann, Vortices behind a bluff body with an upswept aft section in ground effect. Int. J. Heat Fluid Flow **25**(1), 1–9 (2004)
10. Y. Yucheng, L. Wanping, L. Liangjun, Numerical simulation of flow around a bluff body at high Reynolds number. J. Huazhong Univ. Sci. Technol. **31**(2), 106–108 (2003)
11. E. Ulrichs, H. Herwig, Between two limits: flow separation behind a bluff body close to a wall. Forsch. Ing./Eng. Res. **68**(1), 36–38 (2003)
12. N. Tokai, *Visualization of Pressure Field in Turbulent Wake Between Two Bluff Bodies* (Fourth ASME/JSME Joint Fluids Engineering Conf, Honolulu, HI, 2003), pp.6–10
13. M. Rakshantha, S. Sundararaj, B.T. Kannan, Numerical simulation of flow over buildings using OpenFOAM®, in *AIP Conference Proceedings 2019*, vol. 2112, No. 1 (2019) pp. 020149-1

14. A. Smaïli, C. Masson, On the rotor effects upon nacelle anemometry for wind turbines. Wind Eng. **28**(6), 695–714 (2005)
15. A. Smaïli, C. Masson, S.R. Taleb, L. Lamarche, Numerical study of the thermal behaviour of wind turbine nacelle operating in Nordic climate. Numer. Heat Transf. Part B: Fundam. **50**(2), 121–141 (2006)
16. L.D. Tran, C. Masson, A. Smaïli, A stable second-order mass-weighted upwind scheme for unstructured meshes. Int. J. Numer. Meth. FluidsNumer. Meth. Fluids **51**, 749–771 (2006)
17. A. Smaïli, A. Tahi, C Masson, Thermal analysis of wind turbine nacelle operating in Algerian Saharan climate. Energy Procedia **18**, 187–196 (2012)
18. K.N. Tekin, T. Yavuz, E. Koç, Thermal analysis of wind turbine nacelle of 2.5 MW turbines at winter conditions, in *10th International Conference on Heat Transfer, Fluid Mechanics and Thermodynamics* 14–16 July 2014 Orlando, Florida
19. M.A. Mahdi, A. Smaili, Numerical investigations of the thermal behavior of a HAWT nacelle using ANSYS FLUENT. Energy Procedia **1**(141), 394–398 (2017)
20. N. Bhamare, P. Gadhe, Thermal analysis of wind turbine nacelle using OpenFoam. Int. J. Sci. Eng. Res. **10**(7) (2019)
21. M.A. Mahdi, A. Smaili, Y. Saad, Numerical investigations of turbulent natural convection heat transfer within a wind turbine nacelle operating in hot climate. Int. J. Thermal Sci. 147(1), 106143 (2020)
22. Z. Zhang, L. Zhang, Y. Yuan, Numerical simulation of turbulent natural convection heat transfer in an MW-class offshore wind turbine nacelle based on a multi-feature acquisition meshing technique. Sustain. Energy Technol. Assess. 57(1), 103249 (2023)
23. A.S.M. Alkaabi, Numerical in investigation on the cooling of HAWT Nacella. MSc. thesis, The Department of Mechanical Engineering, Karabuk University, Turkey (2022)
24. ANSYS, ANSYS Fluent Theory Guide: Fluent Academic v. 2022R1. Canonsburg, PA
25. J. Sheng, X. Meng, S. Chu, H. Guo, Review of the cooling technology for high-power wind turbines, in *5th International Conference on Advanced Design and Manufacturing Engineering*. Atlantis Press (2015), pp. 1798–1803

Chapter 10
Parameter Interactions on the Adsorption Behaviour of Cobalt onto Saline Soil with Different Biosurfactants

S. Narimannejad, Q. Cai, B. Zhang, K. E. Taylor, and N. Biswas

Abstract This study examined parameter interactions on the adsorption behavior of cobalt(II) (Co) onto saline soil with different biosurfactants using a full factorial design. A 2^4 factorial design was conducted at two levels consisting of 22 runs of possible combinations of four factors, i.e. size of soil particles (A), biosurfactant concentration (B), type of biosurfactant (C) and initial Co concentration (D). The results show that the interaction of ABCD exerts the most significant influence on the remaining Co. The effect of significant terms on response decreases in the order ABCD > C > A and BC > ABC > AC and BCD. Factors A and C have a positive effect on the remaining Co. A higher concentration of remaining Co was observed in the solution when trehalose lipids was applied, compared to surfactin in the same matrix, indicating trehalose lipids solution is a more effective washing agent. Trehalose lipids' high affinity for the soil leads to much more influence of soil particle size on remaining Co when the biosurfactant is surfactin compared with trehalose lipids. The Co concentration remaining in solution is negatively related to the surfactin concentration at the applied initial concentration of Co. Conversely, trehalose lipids had positive impact on the remaining Co concentration in the same matrix. The optimized conditions for the highest Co washing efficiency were determined and the findings illustrate the effects of principal parameter interactions that are used to estimate the mobility of Co in the environment and facilitate the prediction of the associated risks and the remediation strategy.

S. Narimannejad (✉) · Q. Cai · B. Zhang
Faculty of Engineering and Applied Science, Memorial University of Newfoundland, St. John's, NL 1B 3X5, Canada
e-mail: samiran@uwindsor.ca

S. Narimannejad · N. Biswas
Department of Civil and Environmental Engineering, University of Windsor, Windsor, ON N9B 3P4, Canada

K. E. Taylor
Department of Chemistry and Biochemistry, University of Windsor, Windsor, ON N9B 3P4, Canada

D. S-K. Ting and A. Vasel-Be-Hagh (eds.), *Engineering to Adapt*, Springer Proceedings in Energy, https://doi.org/10.1007/978-3-031-47237-4_10

Keywords Adsorption · Cobalt II · Biosurfactant · Parameter interaction · Full factorial design · Saline soil

10.1 Introduction

Cobalt (II) (Co) may be released into the environment from natural and industrial sources [20, 34]. Co occurs naturally in the Earth's crust in carbonate and igneous rocks [4, 16]. It is used in making refractory metals, permanent magnets and pigments as well as being an additive in agriculture, while its alloys have applications in the electrical and aeronautical industries [24, 36]. With industrial growth, the environment is in danger of pollution more than ever before [18, 19, 47]. Ecotoxicity knowledge regarding Co has been established by various organizations using several environmental protection criteria. Co has played a role in the development of diffuse interstitial lung disease and asthma [39]. Although, Co is an essential trace element, high concentrations of it are toxic [35]. Based on cobalt ecotoxicity data, protection criteria have been defined for ecosystems [2, 55]. The toxic dose of cobalt depends on various factors, including the route of exposure (ingestion, inhalation, or skin contact), the chemical form of cobalt, and individual susceptibility. For inhalation exposure to cobalt (dust or fumes), occupational exposure limits are often used as a reference to assess potential toxicity. The Occupational Safety and Health Administration in the United States, for instant, has established a permissible exposure limit of 0.1 mg per cubic meter of air for cobalt and its compounds as an 8-h time-weighted average [52]. For oral exposure, the World Health Organization has established a provisional tolerable weekly intake of 70 μg of cobalt per kilogram of body weight [67]. In animal experiments (e.g., mice), the contamination status of Co was found to be strongly associated with radical activation and lung injury [33]. Exposure to Co in the workplace occurs in industries such as in the manufacture of paints, varnishes, linoleum and in electroplating (Lin et al. 2018). In view of its ecotoxicity and the associated risks in the environment, USEPA [65] has defined guidelines for Co concentration in the environment in order to protect various target organisms. Co, a heavy metal, has considerable toxicity and is also non-degradable with significant bioaccumulation [61, 68]. However, its transport and fate in soils have not been well studied.

An understanding of the behavior of heavy metals in the environment facilitates the prediction of their associated risks and potential remediation strategies. Adsorption is one of the key processes in the fate and transport of heavy metals in the soil [29]. During adsorption, the mobility of a heavy metal depends on its concentration (5–300 mg/L) in the soil, as well as the properties of the metal and soil concerned [17]. Research efforts on Co transferability and mobility in the environment have been very limited. Papelis [53] and Narimannejad et al. [49] found that Co was not highly mobile in soil and its migration depended on soil characteristics. For example, a soil with high clay content was associated with little Co mobility. Further research is needed to fill the knowledge gap.

Several parameters, namely soil particle size/surface area and initial heavy metal concentration play significant roles in metal sorption. The size of soil particles affects the adsorption of heavy metals on the soil, and the size distribution of soil particles affect the transport of heavy metals [40]. The larger the specific surface area (SSA), the finer the soil particles, the stronger is the affinity for heavy metals than for coarse soil particles [7, 23, 28, 69]. However, no studies investigated the effects of soil particle size on the adsorption of Co. Initial concentration of heavy metals also affects their adsorption in soil. The effect of initial Zn(II) concentration on its removal from soil by rhamnolipid has been studied. When the concentration of Zn(II) in solution was raised, Zn(II) adsorption efficiencies of the soil decreased [8]. Aşçı et al. [6] characterized the effect of initial Cd(II) concentration on sorption capacity of soil. The sorption capacities for Ni(II) and Cu(II) onto kaolin were measured [15]. Aşçı et al. [7] loaded the soil with different doses of Cd(II) and used rhamnolipid biosurfactant for the recovery of the heavy metal. The sorbed Cd(II) onto the soil increased with increasing initial Cd(II) concentration [9]. However, rarely did studies report on the impact of initial Co concentration on its adsorption in soil.

Biosurfactants, biologically produced surface-active agents, can potentially be used to influence the sorption of some heavy metals through remediation techniques such as soil washing, or to predict removal efficiencies [45]. Soil washing, or flushing, is a popular technology to remediate heavy-metal contaminated soil as it increases the transport of heavy metals to the liquid phase by desorption [33]. Biosurfactants have been added during soil washing to facilitate the process [44, 62]. They also have advantages of low toxicity, biodegradability, possibility of reuse and very good surface-active properties compared with chemically synthesized surfactants [12]. Surfactants increase heavy metals' mobilization from soil into the liquid phase [60], thus making them easier to recover by washing, pumping or flushing [43]. Aşçı et al. [5] studied the potential of rhamnolipid for the recovery of Cd(II) from kaolin in different concentrations of a biosurfactant (0–100 mM). Cd recovery efficiency increased with increasing rhamnolipid concentration. Wang and Mulligan [66] found that a 0.1% rhamnolipid solution improves heavy metals' removal from mine tailings. After exposing the soil to a 1 mM concentration of Cd(II), 73% of Cd(II) adsorbed onto the soil and 55.9% of the adsorbed Cd(II) was desorbed by an 80 mM solution of rhamnolipid [5]. Mulligan et al. [45] used rhamnolipid, sophorolipid and surfactin to remove Cu(II) and Zn(II) from metal-contaminated sediment and found 65%, 25% and 15% of the Cu(II) and 18%, 60% and 6% of the Zn(II) were removed, respectively. Hong et al. [27] used saponin to effectively enhance recovery of heavy metals from soils, showing 90–100% of Cd(II) removal and 85–98% of Zn(II) removal. [21] investigated the adsorption and desorption characteristics of Cd(II), Zn(II) and Cu(II) ions on/from digested sludge. They studied the effect of type and dose of biosurfactant on desorption with saponin and rhamnolipid JBR 515 at initial concentrations of 0–250 g/L. Singh and Cameotra [63] studied the application of lipopeptide biosurfactants including surfactin and fengycin at different concentrations (i.e., 0, 0.5, 1, 10, and 50 CMC (critical micelle concentration)) to remove heavy metals (i.e., Fe, Pb, Ni, Cd, Cu, Co and Zn) from the soil. From these studies, it was found that the type and dosage of biosurfactants have significant impacts on the mobility of heavy metals. Narimannejad et al. [49] studied the impact of different concentrations (1 and 2 CMC) of each biosurfactant including surfactin, trehalose lipids and rhamnolipid on Co adsorption onto soil at different initial concentrations of Co (i.e., 50, 100, 200, 300, 400 mg/L). High affinity of Co for the soil was reported even in the presence of the biosurfactants. However, very little information is available concerning the influence of a biosurfactant on the Co adsorption process on the soil. Additionally, as all of these studies investigating the desorption behavior of the accumulated heavy metals generally differ from that of adsorption [9], there is a knowledge gap in heavy metal adsorption behavior (especially Co) on the soil with varied types and dosages of biosurfactants.

There may be interactions among soil particle size, biosurfactant concentration, type of biosurfactant and initial concentration of heavy metals. The existence of interactions complicates an understanding of the individual effects. For example, soil with a fine particle size (such as clay) has a complex impact on heavy metals' sorption/desorption in soil when interacting with biosurfactants at varied concentrations [8, 22], but this has not been well studied [22]. The existing studies working on the abovementioned 4 factors with heavy metals sorption in soil have been focused on individual effects. No study considered them and their interactions in a systematic manner using a mathematically and statistically supported approach; instead, a one-factor-at-a-time approach (OFAT) has been designed to analyze effects. OFAT is unable to estimate interactions, which can be significant in this scenario [3]. To explore the interactions between operating factors, statistical design for experiments (DOE) (e.g., response-surface methodology) has been used for modeling and optimization [32]. This method is widely known in industry to improve performance of products and processes. This mathematical and statistical approach saves experiment run times compared with OFAT [26]. As one type of response-surface methodology, factorial design is a well-established approach to study interactions between factors [3, 41]. In such statistical design experiments, adequate data can be derived with a minimum number of experimental runs, because the factors are simultaneously changed at their levels [59]. In the full-factorial experiments, not only the effects of more than one factor on results are investigated, but also their relative significance in the process is obtained and the interactional impacts of two or more factors can also be investigated [57]. No study thus far has attempted to use DOE methodology to investigate how these four factors, size of soil particles, biosurfactant concentration, type of biosurfactant and initial Co concentration, affect cobalt adsorption in soil.

Moreover, the sorption can be affected by various environmental conditions such as soil salinity [1, 7]. It was found that increasing soil salinity led to higher heavy metal mobility, and the extent of increased mobility varied with the type of heavy metal [1, 25]. However, no studies have investigated the effects of soil salinity on Co adsorption in the soil, despite the prevalence of saline or sodic soils with shallow saline or sodic groundwater levels [58].

Therefore, to fill the knowledge gap and tackle the challenge, this study reports on a 2^4 factorial design to examine the main effects and quantify interactions among soil-particle size, biosurfactant concentration, type of biosurfactant and initial Co concentration on the adsorption of cobalt in saline soil to determine optimized conditions. The results obtained facilitate an understanding of cobalt's transport and fate in the subsurface environment and their potential associated risks. Moreover, by understanding the environmental behavior of Co in the soil, the appropriate methods for soil remediation can be determined.

10.2 Materials and Methods

10.2.1 Soil Preparation

To assess adsorption of Co, batch equilibrium experiments were performed using soil comprised of 50% natural clay and 50% sand. This mixture was chosen simply to make the sample more manageable while still having the properties of clay. The soil was dried at 105 °C overnight followed by the removal of gravel and rocks. At first, sand was sieved separately to pass through a 2 mm stainless steel mesh, and then a fraction of sieved sand was sieved again through a 0.59 mm stainless steel mesh. To sieve the clay, only 0.59 mm mesh was used. For the two soil samples used in this study, equal weights of the sieved fractions were combined, one with <0.59 mm clay with <0.59 mm sand, and the second with <0.59 mm clay with 0.59–2.0 mm sand. Then for each soil sample, 0.087 g NaCl was added to obtain 3% saline soil, to mimic the properties of coastal soil. In the research conducted by [49], soil properties, including pH, soil bulk density, soil water content, soil cation exchange capacity and major element oxides as well as trace element concentrations in the soil were characterized.

10.2.2 Biosurfactant Production

Two different biosurfactants were prepared in the Northern Region Persistent Organic Pollution Control (NRPOP) Laboratory at Memorial University of Newfoundland. The first biosurfactant, surfactin, was produced by *Bacillus subtilis* N3-1P isolated from oily seawater [12]. This product is a lipopeptide biosurfactant which can reduce water surface tension significantly [70]. The second biosurfactant, from a hyper-production mutant, *Rhodococcus erythropolis* M36, was generated through three runs of UV mutagenesis and productivity screening of a wild strain isolated from produced water samples from offshore Newfoundland, Canada [11]. This biosurfactant was identified as glycolipids (trehalose lipids). Surfactin and trehalose lipids were prepared by the protocols of [70] and [12], respectively.

From previous experiments by [49], the powdered surfactin and the viscous syrup trehalose lipids products had critical micelle concentrations (CMC) of 0.09 and 2.6 g/L and surface tensions of 32.5 and 33 mN/m, respectively. Surface tension was measured by the ring method using a Du Nouy Tensiometer. Surface tensions of the biosurfactant as a function of concentration of biosurfactant were plotted. The CMC then was obtained from the intercept of two straight lines extrapolated from the concentration-dependent and concentration-independent parts.

10.2.3 Co Adsorption onto Soil with Biosurfactant Addition

Co adsorption onto saline soil with various initial Co concentrations, dose/type of biosurfactants and size of soil particles was conducted. Each 2.9 g sample of dry, homogenized, 3% saline soil was equilibrated in a polypropylene centrifuge tube with 50 mL of Co solution (100, 200 or 300 mg/L) followed by the addition of a biosurfactant. The biosurfactants (surfactin or trehalose lipids) were applied at three concentrations (0.5, 1 and 1.5 CMC). Three different sand-particle sizes, (1) 2.0–0.59 mm, (2) $\leq$0.59 mm, and (3) 1:1 mixture of (1) and (2), were used in the adsorption process. The clay, mixed in equal weight with each sand fraction, used was $\leq$0.59 mm.

A pH of 9 and reaction time of 4 h for conducting the Co adsorption onto the soil were selected based on a previous one-factor-at-time (OFAT) study by [49]. Samples before adsorption experiments were adjusted to pH 9 by 1 M HCI or 1 M NaOH solution. During adsorption, the pH level of each run was kept constant by adding small volumes of 0.1 M HCI or 0.1 M NaOH solution. Co adsorption was performed by shaking each tube in a shaker at 200 rpm at room temperature for 4 h, at which time the tube was centrifuged at 6,000 rpm for 30 min. The aliquots of the supernatant solution were acidified to pH 2.0 using 2% HNO_3 for flame atomic absorption spectroscopy (SpectrAA, Varian Australia Pty Ltd) measurement in triplicate to determine the Co concentration remaining in the solution after each adsorption treatment. Liquid samples were diluted, and Co was detected through the characteristic wavelength from a Co lamp. A calibration curve for Co was constructed using standard solutions in the range of 0–7 mg/L.

10.2.4 Experimental Design

DESIGN EXPERT 8.0.6 statistical software (Stat-Ease, Inc. Minneapolis, USA) was employed for experimental design and statistical analysis. The 2^4 factorial experiment was conducted at two levels, a high and a low level. The lower level of factors was coded as '-1' and the higher level was coded as '+1'. The full factorial design resulted in 22 runs (16 run combinations plus 6 center points). The mathematical equation in terms of coded factors can be expressed as:

$$Y = a_0 + a_1 A + a_2 B + a_3 C + a_d D$$

$$+ a_5AB + a_6AC + a_7AD + a_8BC + a_9BD + a_{10}CD + a_{11}ABC + a_{12}ABD + a_{13}ACD + a_{14}BCD + a_{15}ABCD \quad (10.1)$$

where Y is the theoretical yield function as mg/L Co remaining in solution. The a_0, a_{1-4} and a_{5-15} represent the global mean and the regression coefficients corresponding to the main factor effects and interactions, respectively.

These 22 experimental runs were accomplished by varying the four factors at -1 and +1 levels plus triplicate center points which are entered as (0) in Table 10.2 below for each numerical factor. The factors and levels include the size of sand particles (≤0.59 mm and between 2 and 0.59 mm), biosurfactant concentration (0.5 CMC and 1.5 CMC), type of biosurfactant (surfactin and trehalose lipids) and initial Co concentration (100 and 300 mg/L) (Table 10.1). The levels of design were selected based on results from our previous OFAT study by [49]. The response variable is the concentration of the remaining Co in the solution. Table 10.2 lists the experimental layout of the factorial design.

10.3 Results and Discussion

10.3.1 Development of Regression Model Equation

The significance of the factors affecting the Co adsorption onto the soil were determined by performing an analysis of variance (ANOVA) shown in Table 10.3. ANOVA is generally conducted to survey the reliability of the model [31]. It illustrates whether the variations obtained from a model are significant compared to the variations related to experimental error [30]. A sum of squares (SS) of each factor shows its significance in the process. The greater the SS value, the more important is the factor in the experiencing process. The p-value of the model ($0.0345 < 0.05$) and lack of fit ($0.4082 > 0.05$) indicate the model is significant and adequate. The main effects and interactions of each factor with p-values less than 0.05 are considered as statistically significant. The size of particle-biosurfactant concentration-type of biosurfactant-initial Co concentration interaction (ABCD) has the greatest effect on the residual concentration of Co in the solution, followed by the type of biosurfactant (C). Among

Table 10.1 Factors and levels selected for factorial design

Factor	Type	Code	Low (−1)	High (+1)
Size of particle (mm)	Numeric	A	≤0.59	0.59–2
Biosurfactant concentration (CMC)	Numeric	B	0.5	1.5
Type of biosurfactant	Categorical	C	Surfactin	Trehalose lipids
Initial Co concentration (mg/L)	Numeric	D	100	300

Table 10.2 Experimental layout (coded) of the factorial design

Standard order	Run	Factor A	Factor B	Factor C	Factor D
		Particle size	Biosurfactant concentration	Type of biosurfactant	Initial Co concentration
1	1	−1	−1	Surfactin	−1
5	2	−1	−1	Trehalose lipids	−1
22	3	0	0	Trehalose lipids	0
18	4	0	0	Trehalose lipids	0
3	5	−1	1	Surfactin	−1
19	6	0	0	Surfactin	0
15	7	−1	1	Trehalose lipids	1
20	8	0	0	Trehalose lipids	0
6	9	1	−1	Trehalose lipids	−1
14	10	1	−1	Trehalose lipids	1
17	11	0	0	Surfactin	0
2	12	1	−1	Surfactin	−1
16	13	1	1	Trehalose lipids	1
9	14	−1	−1	Surfactin	1
12	15	1	1	Surfactin	1
21	16	0	0	Surfactin	0
11	17	−1	1	Surfactin	1
10	18	1	−1	Surfactin	1
13	19	−1	−1	Trehalose lipids	1
4	20	1	1	Surfactin	−1
7	21	−1	1	Trehalose lipids	−1
8	22	1	1	Trehalose lipids	−1

the significant terms, size of particle (A) and biosurfactant concentration-type of biosurfactant interaction (BC) have the lowest effect, with p-values equal to 0.0429.

The equation in terms of coded factors is given in Eq. (10.2) after substituting the regression coefficients by their numerical values. For the categorical Factor C, value of -1 is used for surfactin and $+1$ for Trehalose lipids.

$$\begin{aligned}\text{Final Co concentration} = {} & 1.16 + 0.14\text{A} + 0.081\text{B} + 0.14\text{C} \\ & + 0.056\text{D} - 0.044\text{AB} - 0.12\text{AC} - 0.094\text{AD} \\ & + 0.14\text{BC} - 0.056\text{D} + 0.019\text{CD} - 0.13\text{ABC} \\ & - 0.056\text{ABD} + 0.069\text{ACD} - 0.12\text{BCD} - 0.27\text{ABCD}\end{aligned} \tag{10.2}$$

The coefficient of determination of the regression model (R^2) implies that 92.02% of the variations in the remaining Co can be explained by the model. Adjusted R^2 is equal to 72.08% which means that the model is fit enough for current observation. However, the predicted R^2 obtained (-5.8477) implies the current model is not a good predictor of the response. By removing the insignificant terms from the model, predicted R^2 would be improved. However, as factor ABCD is significant, removing any other factors would compromise the hierarchy of the model, for example in Table 10.3 interactions AC and BCD have p-values greater than 0.05 but are kept in Table 10.4 to prevent compromising hierarchy of the model. The hierarchy of the

Table 10.3 Significance of factor effects on the response variable

Source	Sum of squares	df	Mean square	F-value	p-value
Model	3.49	15	0.23	4.61	0.0345
A	0.33	1	0.33	6.56	0.0429
B	0.11	1	0.11	2.09	0.1980
C	0.44	1	0.44	8.66	0.0258
D	0.051	1	0.051	1.00	0.3551
AB	0.031	1	0.031	0.61	0.4655
AC	0.23	1	0.23	4.47	0.0788
AD	0.14	1	0.14	2.79	0.146
BC	0.33	1	0.33	6.56	0.0429
BD	0.051	1	0.051	1.00	0.3551
CD	5.63E−03	1	5.625E−003	0.11	0.7498
ABC	0.28	1	0.28	5.46	0.0580
ABD	0.051	1	0.051	1.00	0.3551
ACD	0.076	1	0.076	1.50	0.2667
BCD	0.23	1	0.23	4.47	0.0788
ABCD	1.16	1	1.16	22.91	0.0030
Lack of fit	0.11	2	0.055	1.13	0.4082

model is used to ensure proper transition between coded-factor models and actual-factor models. In the present case, the model is only expressed in the coded factors to eliminate the dependency on model hierarchy. Therefore, the insignificant terms of the model can be removed to improve the model predictability and make adjusted R^2 and predicted R^2 in agreement, with difference less than 0.25 [13, 42].

Table 10.4 shows the new analysis of variance (ANOVA) after removing insignificant factors, revealing that the model has an insignificant curvature with F-value of 0.93 and p-value of 0.4202. Therefore, the model shown in Table 10.4 is not adjusted for curvature. The p-value of the model ($0.0008 < 0.05$) and lack of fit ($0.4372 > 0.05$) indicate the model is significant with good fit. The main effects and interactions of each factor with p-values less than α-level are considered as significant. With more than 95% confidence, interaction of ABCD (p-value = 0.0005) has the greatest effect on the residual concentration of Co in the solution, followed by the type of biosurfactant (factor C) with p-value = 0.0159. The size of particle (A), biosurfactant concentration-type of biosurfactant interaction (BC), with p-values of 0.0317, and size of particle-biosurfactant concentration-type of biosurfactant interaction ($\text{p-value}_{ABC} = 0.0469$) also have significant effect on the final Co concentration. With more than 90% confidence, size of particle-type of biosurfactant interaction (AC) and biosurfactant concentration-type of biosurfactant-initial Co concentration interaction (BCD), with p-values equal to 0.0688, have the lowest effect on the final Co concentration.

The reduced model has R^2 of 0.7858 which shows 78.58% of variability in the response is explained by the variables noted. The predicted R^2 of 0.42 is in reasonable agreement with the adjusted R^2 of 0.67 (difference < 0.25). Adequate precision is a measure of the signal-to-noise ratio and the favorable value of 9.597 (>the desirable value 4) indicates an adequate signal [64]. Coefficient of Variation (C.V. %) is the error expressed as a percentage of the mean, the lower the value of, the more precise the estimate. The coefficient of variation of 20.79 shows the deviations between predicted and actual values and affirms the reliability and precision of the conducted

Table 10.4 Significance of factor effects on the response variable after removal of insignificant factors

Source	Sum of squares	df	Mean square	F-value	p-value
Model	2.98	7	0.43	7.34	0.0008
A	0.33	1	0.33	5.70	0.0317
C	0.44	1	0.44	7.53	0.0159
AC	0.23	1	0.23	3.89	0.0688
BC	0.33	1	0.33	5.70	0.0317
ABC	0.28	1	0.28	4.75	0.0469
BCD	0.23	1	0.23	3.89	0.0688
ABCD	1.16	1	1.16	19.91	0.0005
Lack of fit	0.62	10	0.062	1.28	0.4372

experiments [56]. PRESS is a measure of how well a particular model fits each point in the design. PRESS can be used to evaluate the performances of various models, a lower value indicates better structure of the candidate models. The reduced model PRESS value of 2.22 shows dramatic improvement compared to the full model PRESS value (25.97). The reduced model has lower PRESS, hence higher predicted R^2 indicating it has lower error and structure of the model is significantly improved [13].

After the reduction, the regression model in terms of coded factors is given in Eq. (10.3) after substituting the regression coefficients by their numerical values as given in Table 10.5. Percent contributions are calculated by the software (DESIGN EXPERT) for the significant factors according to their p-values; the percent contributions do not sum to 100% because some factors have been discarded from Table 10.3.

$$\begin{aligned}\text{Final Co concentration} &= 1.16 + 0.14\text{A} + 0.14\text{C} - 0.12\text{AC} + 0.14\text{BC} \\ &\quad - 0.13\text{ABC} - 0.12\text{BCD} - 0.27\text{ABCD}\end{aligned} \tag{10.3}$$

The diagnostic graphs for the ANOVA and the model are shown in Fig. 10.1. Figure 10.1a shows the Half-Normal plot which sheds light on the importance of each factor. The further away the terms appear from the diagonal line, the more important they are. Figure 10.1b shows an agreement between the predicted and corresponding experimental results. The normal probability plot of residual values is shown in Fig. 10.1c, to determine whether the data set is normally distributed. If the points fall close to the straight line, it means that the data set has normal distribution. For the statistical study, it is necessary to assume that the data come from a normal distribution [30]. It is evident from Fig. 10.1c, that the points are well distributed along the inclined line and no pattern is seen. In Fig. 10.1d (predicted values vs. internally studentized residuals) the points are randomly scattered. No point is beyond the borders and the residuals have constant variance. As shown in Fig. 10.1e (run number vs. internally studentized residuals) no trend is seen, which

Table 10.5 Estimated regression coefficients of factors and their effects on remaining Co in the solution after adsorption

Term	Effect	Coefficient estimate (coded)	% Contribution
Constant	–	1.16	–
A	0.29	0.14	8.61
C	0.33	0.14	11.38
AC	−0.24	−0.12	5.88
BC	0.29	0.14	8.61
ABC	−0.26	−0.13	7.18
BCD	−0.24	−0.12	5.88
ABCD	−0.54	−0.27	30.10

leads to an independence of observations assumption. Thus, ANOVA assumptions are valid and the adequacy of the model is confirmed.

The experimental and predicted remaining Co values are presented in Table 10.6 in coded form where the coding is explained in Table 10.1. The residual values show the difference between the predicted and observed values. The absolute values of the residuals are small and within the range of 0–0.48 mg/L.

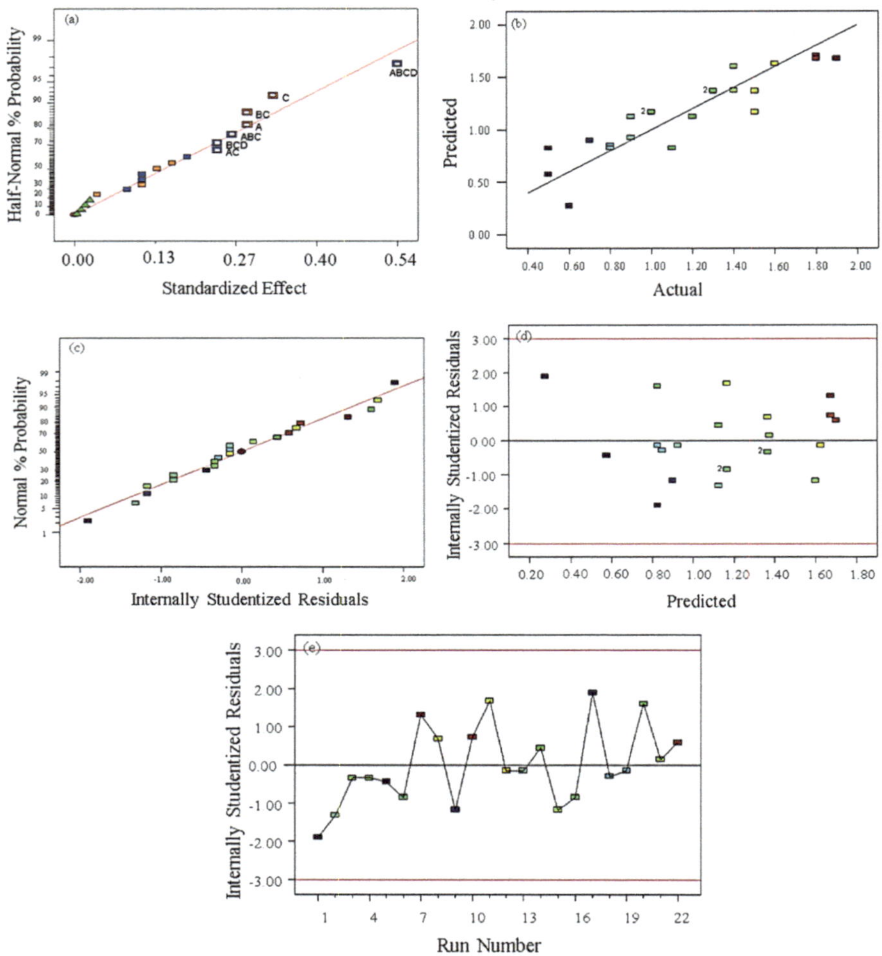

Fig. 10.1 Diagnostic graphs for ANOVA **a** half-normal plot, **b** predicted versus actual plot, **c** normal plot of residuals, **d** residuals versus predicted plot, and **e** residuals versus run plot

Table 10.6 Factorial design matrix of four factors along with predicted and experimental results for remaining Co in the solution after adsorption

# Run	Coded values of independent factors					Remaining Co (mg/L)		
	A	B	C	D	Observed	%RSD	Predicted	Residual
1	−1	−1	−1	−1	0.5	0.3	0.88	−0.38
2	−1	−1	+1	−1	0.9	0.3	1.15	−0.25
3	0	0	+1	0	1.3	0.9	1.3	0.0
4	0	0	+1	0	1.5	2.4	1.3	0.2
5	−1	+1	−1	−1	0.5	0.4	0.63	−0.13
6	0	0	−1	0	1.5	1.6	1.02	0.48
7	−1	+1	+1	+1	1.9	1.3	1.7	0.2
8	0	0	+1	0	1.3	0.9	1.3	0.0
9	+1	−1	+1	−1	0.7	1.6	0.93	−0.23
10	+1	−1	+1	+1	1.8	0.3	1.7	0.1
11	0	0	−1	0	1	1.5	1.02	−0.02
12	+1	−1	−1	−1	1.6	0.6	1.68	−0.08
13	+1	+1	+1	+1	0.9	0.4	0.95	−0.05
14	−1	−1	−1	+1	1.2	1.1	1.18	0.02
15	+1	+1	−1	+1	1.4	0.6	1.66	−0.26
16	0	0	−1	0	1	0.9	1.02	−0.02
17	−1	+1	−1	+1	0.6	0.3	0.33	0.27
18	+1	−1	−1	+1	0.8	0.8	0.91	−0.11
19	−1	−1	+1	+1	0.8	1.0	0.85	−0.05
20	+1	+1	−1	−1	1.1	1.1	0.88	0.22
21	−1	+1	+1	−1	1.4	1.3	1.4	0.0
22	+1	+1	+1	−1	1.8	2.4	1.73	0.08

10.3.2 *Quantitative Estimation Effects of the Factors*

Figure 10.2 shows the main effect of the significant numerical factor A (size of particles) at a low level of categorical factor C (surfactin) on Co adsorption in the saline soil. According to Eq. (10.3), the positive sign of the coefficient of factor A (0.29 in Table 10.5) means that size of particle has a positive impact on remaining Co in the solution. In Fig. 10.2, increasing the size of particle from 0.59 to 2 mm increases the remaining Co from 0.76 to 1.3 mg/L in the solution. This is due to the decrease of the surface area of the soil with bigger soil particles.

Figure 10.3 shows the main effect of factor A on Co adsorption in saline soil when trehalose lipids were used. As seen, the Co remaining was essentially insensitive to particle size, in contrast to Fig. 10.2 for surfactin which shows marked particle-size

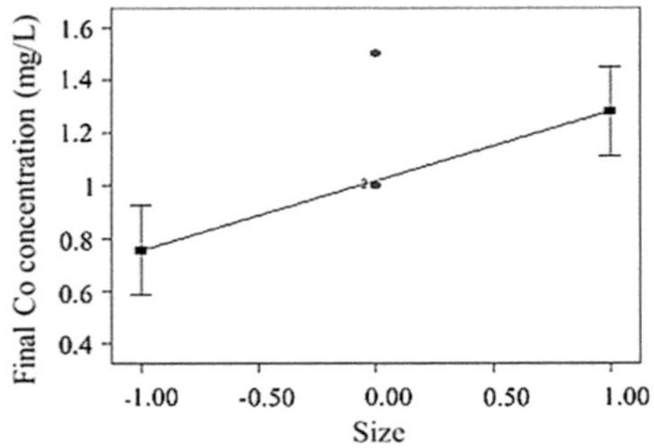

Fig. 10.2 Main effect plot for remaining Co concentration with surfactin

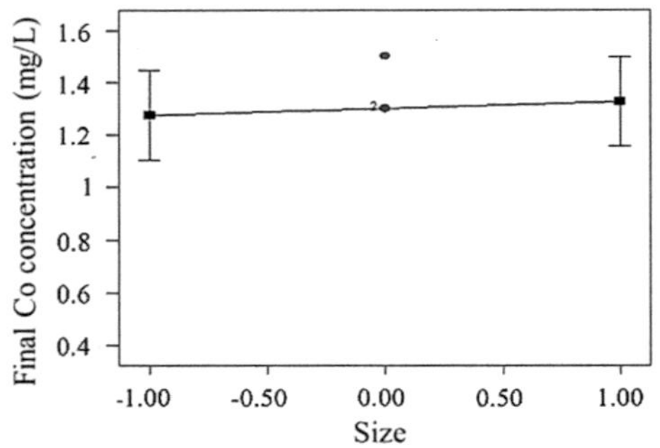

Fig. 10.3 Main effect plot for remaining Co concentration with trehalose lipids

sensitivity. Thus, type of biosurfactant (factor C, 0.33 in Table 10.5) has a significant impact on the remaining Co.

Figure 10.4 shows the Co remaining in the solution when the two biosurfactants are compared at their CMCs, final Co concentrations are 1.0 and 1.3 mg/L, respectively. With trehalose lipids, more remaining Co may be observed due to: (1) the high tendency of trehalose lipids being adsorbed by the soil, according to the reported measurements of surface tension changes in the solution by [48] and [49], which leads to the decrease of available soil surface area for Co adsorption, (2) the different degrees of complexation between Co and the two biosurfactants [51].

10.3.3 Evaluation of the Effects of Parameter Interactions on Co Adsorption

The residual Co found in the solutions after adsorption is mostly dependent on the negative interaction among ABCD (−0.54 in Table 10.5). Furthermore, along with the strong effects of particle size and surfactant type noted above, the interaction effect between surfactant concentration (factor B) and surfactant type (factor C), (0.29 in Table 10.5) is also strong. Somewhat smaller effects are shown among size,

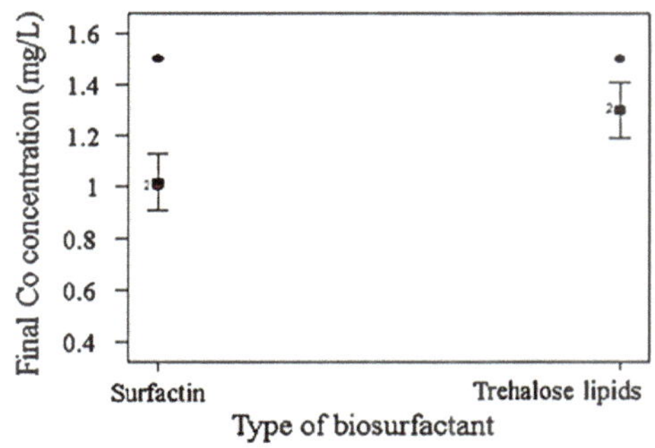

Fig. 10.4 Main effect plot of biosurfactant type (at CMC) on Co remaining

surfactant concentration and type (−0.26 for ABC in Table 10.5), between size and type (−0.24 for AC in Table 10.5) and among surfactant concentration, type and Co concentration (−0.24 for BCD in Table 10.5). The p-values from ANOVA in Table 10.4 confirm the significance of the terms in Table 10.5.

Figure 10.5 shows two-factor interaction between the type of biosurfactant (C) and the size of soil particle (A). Interaction of AC displays that an increased size of the soil particles results in an increase of the remaining Co concentration in the solution when the each biosurfactant is applied. As illustrated in Fig. 10.5, by considering interaction of AC, the influence of the size of soil particles on the result is stronger when the applied biosurfactant is surfactin (from 0.76 to 1.3 mg/L) compared to trehalose lipids applied (from 1.28 to 1.33 mg/L), which can be explained by high tendency of trehalose lipids to be adsorbed to the soil particles [48]. When the surface of the particles is occupied with trehalose lipids, consequently there is not enough space for Co adsorption and will therefore result in the high remaining Co in the solution at each particle-size of the soil.

Figure 10.6 shows two-factor interaction (BC) between the type of biosurfactant (C) and the concentration of each biosurfactant (B). Interaction of BC illustrates that increased concentrations of surfactin and trehalose lipids result in a decrease (from 1.2 to 0.9 mg/L) and an increase (from 1.2 to 1.4 mg/L) of the remaining Co in solution, respectively.

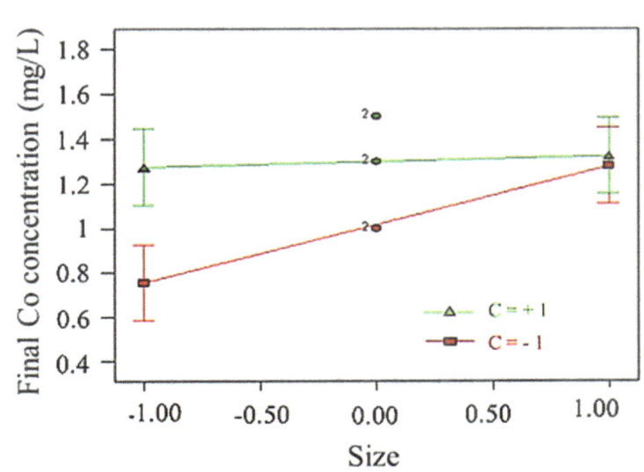

Fig. 10.5 Interaction effect of surfactant type and size of soil particles (AC) on the response variable remaining Co

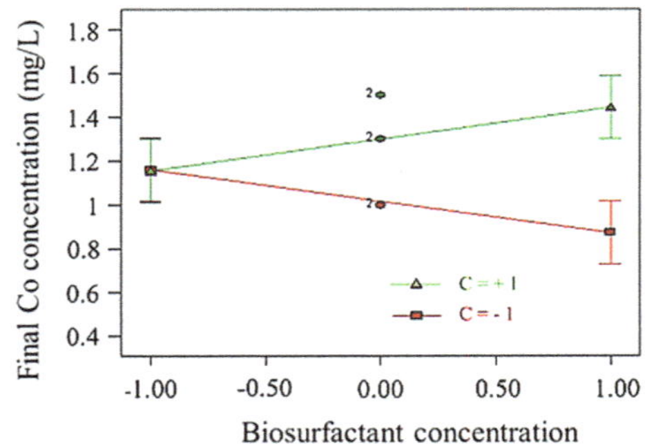

Fig. 10.6 Interaction effect of biosurfactant (−1, surfactin; +1, trehalose lipids) type and biosurfactant concentration (BC) on the response variable remaining Co

A closer examination of this interaction leads to interpretation of the interactions among Co, soil particle size and biosurfactants. In the literature, three modes of interaction have been proposed, competition, binding/complexing and biosurfactant assembly (e.g., micelle, vesicle and lamellar) entrapment [5, 7, 9, 14, 43, 51, 63]. Firstly, heavy metals, as cations, adsorb on the soil particles owing to the negative charge of the soil particle surface [9]. Some biosurfactants have also been found to have good soil affinity because of their hydrophobicity [7, 9, 51, 63]. The introduction of such biosurfactants can serve as competitors to the Co-soil adsorption. Secondly, some biosurfactants have been found to form complex and/or ionic bonds with Co and affect Co-soil adsorption [51]. Lastly, when biosurfactants are added above their critical micellar concentrations, they can form assemblies including micelles, vesicles, and lamellar phases [7, 51]. These assemblies have been found to entrap heavy metals in previous studies [63]. When micelles are formed with negatively charged biosurfactants, they can also bind to Co as well [9]. Trehalose lipids is a negatively charged material and surfactin has both positive and negative charge sites [10]. The size and shape of assemblies vary but they are generally large enough to be retained by ultrafiltration [38]. Assemblies may be entrapped in the voids of soil particles, depending on the size of the soil particles [7]. As expected, the types of biosurfactants affect their soil affinity, binding/complexing ability to Co, water solubility [63] and the size and shape of micelles, and will therefore affect all three modes of interaction. Surfactin has relatively higher water solubility than the trehalose lipids [10]. When surfactin is used with an increased concentration, more competition and complexing/binding to soil particles are expected, resulting in a higher remaining Co concentration in the solution. However, as shown in Fig. 10.6, the opposite holds true. This may be explained by the formation of micelles at 1.5 CMC, leading to the entrapment of Co containing micelles in the soil voids, and the impact of this mode of interaction overpowers the other two modes. In the case of trehalose lipids, with an increased concentration, a higher concentration of remaining Co can be found in the solution (Fig. 10.6). This may be explained by the more significant effects of the first two modes of interaction, rather than the micelle formation.

Figure 10.7 shows interaction of soil-particle size and biosurfactant concentration (AB) for the two surfactants. Figure 10.7a illustrates that larger particle size results in higher remaining Co concentration in the solution at either 0.5 CMC or 1.5 CMC of surfactin, presumably due to decreased surface area for Co adsorption. At the small soil-particle size, low surfactin concentration leads to higher solution Co concentration (1.03 mg/L) than high surfactin concentration (0.48 mg/L). A rationale for this result is that, although the small particle size presents enough surface area for Co binding, surfactin at a concentration below CMC complexes with either or both Co and soil, leaving more Co in the solution phase; by contrast, surfactin in micellar form complexes Co and those 'loaded' micelles bind to the soil, resulting in a lower solution Co concentration. Table 10.7 summarizes the trends under above-mentioned conditions.

In Fig. 10.7b, interaction of AB shows that an increased size of the soil particles results in an increase and decrease of the remaining Co concentration in the solution at low and high concentration of trehalose lipids, respectively. At small particle size,

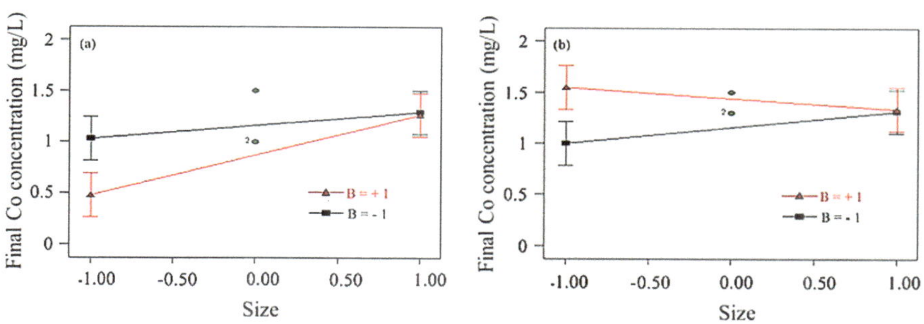

Fig. 10.7 Interaction effect of AB (size of soil particles and biosurfactant concentration) and surfactant type. **a** Surfactin and **b** trehalose lipids on the response variable remaining Co

Table 10.7 Trends in interaction effect of AB (size of soil particles and biosurfactant concentration) and C (type of biosurfactant) on remaining Co

Type of biosurfactant	Trend in solution Co			
	Large particle size		Small particle size	
	0.5 CMC	1.5 CMC	0.5 CMC	1.5 CMC
Surfactin	Decreased soil surface area → high remaining Co		Lack of surfactin for complex formation high remaining Co	Co-loaded micelles adsorbed to the soil low remaining Co
Trehalose lipids	Lack of trehalose lipids for complex formation high remaining Co	More complex formation on the soil low remaining Co	Low dose of trehalose lipids to adsorb to the soil low remaining Co	High soil affinity of trehalose lipids high remaining Co

Co remaining in the presence of high and low trehalose lipids concentration is 1.6 mg/L and 1.0 mg/L, respectively. Trehalose lipids' high soil affinity is consistent with a higher Co concentration with a high concentration of the biosurfactant than that with the low biosurfactant concentration and small soil particles. Thus, with small soil particles, in competition among trehalose lipids, Co and/or Co-trehalose lipids complex to be adsorbed to the soil, the affinity of trehalose lipids is the highest, resulting in a high Co concentration in the solution.

Available soil surface area (inversely proportional to particle size) affects its physical and chemical properties [54] which, in turn, affect soil's ability to sorb contaminants, and the efficiency of soil remediation methods [50]. Thus, in this work, by decreasing the soil surface area (large particles), a Co-trehalose lipids complex wins the competition to be adsorbed to the soil. When trehalose lipids is used with large soil particles and the low concentration of biosurfactant, resulting in a decreased surface area of the soil and not forming the high amount of Co-trehalose complex, respectively, the Co concentration remaining in solution is increased.

When high concentration of trehalose lipids is applied at the large soil-particle size, a large amount of Co-trehalose lipids complex is formed to be adsorbed to the soil, thus, remaining Co concentration in the solution is decreased. Table 10.7 summarizes the trends in the interaction of AB and trehalose lipids.

It can also be observed from Fig. 10.7 that with large soil particles Co remaining in solution was essentially independent of surfactant concentration and type. Therefore, with the large soil particles, sub-CMC concentrations of the biosurfactants would be more cost effective. This may indicate that the applied high concentration of biosurfactants, at large soil particles, is not enough to interact with the existing amount of Co in the soil system [9, 51].

The response of certain variables is a function that explains how the response surface changes as the level of the variables move, when the other variables are fixed at their center points. This function can be shown using contour or surface plots. The contour plots and surface plots for interactions between numerical factors are shown in Figs. 10.8, 10.9, 10.10, and 10.11. The interactions of the size of particle and biosurfactant concentration (AB) at both low (Figs. 10.8, 10.9, 10.10, and 10.11b) and high (Figs. 10.8, 10.9, 10.10, and 10.11c) initial concentrations of Co (D) are greater than at center point (Figs. 10.8, 10.9, 10.10, and 10.11c) of initial Co concentration (D) for either biosurfactant.

It can also be observed from Figs. 10.8, 10.9, 10.10, and 10.11c that when the initial Co concentration was high, the impact of the particle size and concentration of the biosurfactant (AB) on the remaining Co would not be much stronger than that at a lower initial Co concentration. This may show that the applied biosurfactant is not enough to interact with the existing amount of Co in the soil system [9, 51].

Fig. 10.113D Surface for interaction effects for remaining Co at a center point (200 mg/L), b low level (100 mg/L) and c high level (300 mg/L) of initial Co concentration with trehalose lipids

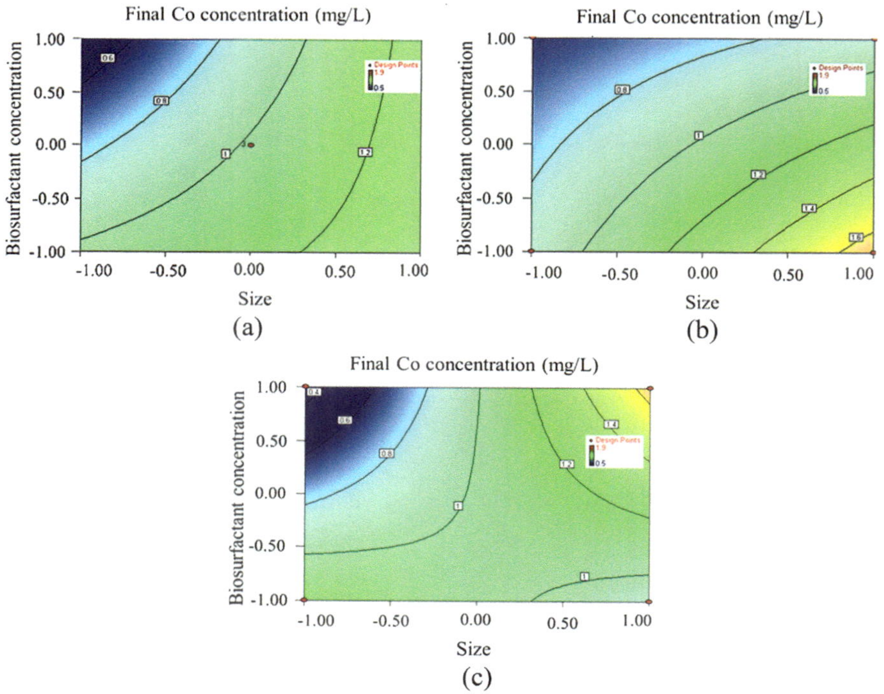

Fig. 10.8 The contours of interactive graphs for remaining Co at **a** center point (200 mg/L), **b** low level (100 mg/L) and **c** high level (300 mg/L) of initial Co concentration with surfactin

10.3.4 Process Optimization, Limitations and Future Work

Based on the current model, the optimized conditions can be obtained using the method described by [46], which optimizes removal from an initial Co concentration of 100 mg/L, the 1.5 CMC of trehalose lipids, and the size particle of $\leq$2 mm. Under these conditions, the remaining Co reaches 1.7 mg/L and desirability is equal to 0.875.

Figure 10.12 shows that at a low level of D and a high level of B and A give the maximum yield when trehalose lipids is applied. When surfactin is used, the highest yield occurs with factors B and D are at their low levels and factor A is at the maximum level (strength is equal to 1.681).

This study attempts to examine the effects of parameter interactions on the adsorption behaviour of cobalt(II) onto saline soil with different biosurfactants within the range of the set levels. Using a numerical model to effectively predict and optimize the processes will be the objective of our future research by incorporating factors of wider range and a systematic model validation.

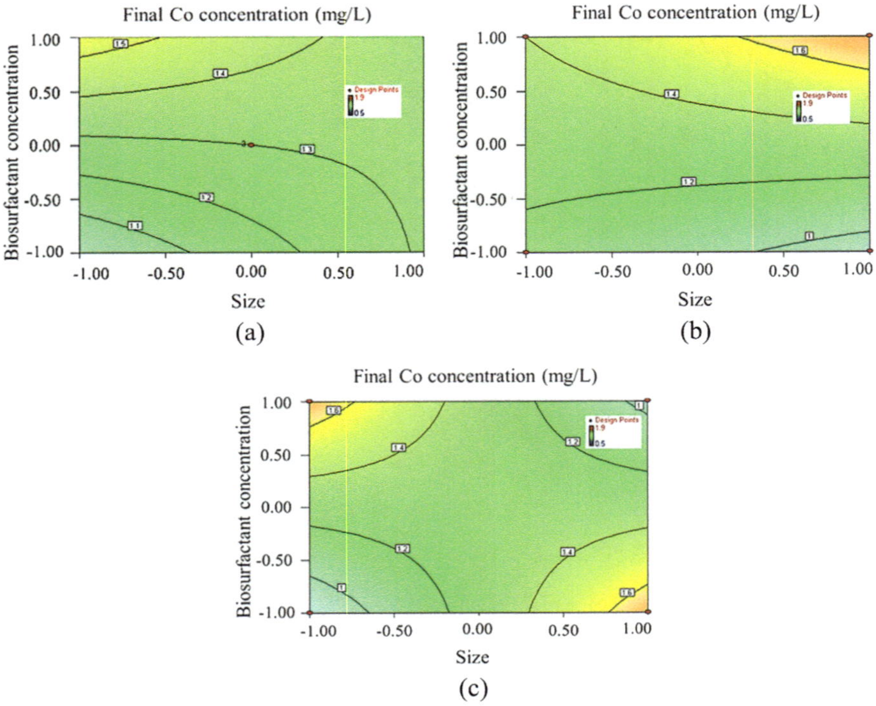

Fig. 10.9 The contours of interactive graphs for remaining Co at **a** center point (200 mg/L), **b** low level (100 mg/L) and **c** high level (300 mg/L) of initial Co concentration with trehalose lipids

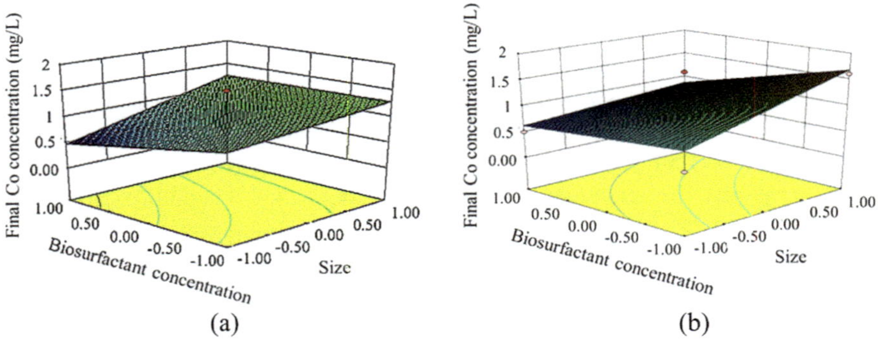

Fig. 10.10 3D Surface for interaction effects for remaining Co at **a** center point (200 mg/L), **b** low level (100 mg/L) and **c** high level (300 mg/L) of initial Co concentration with surfactin

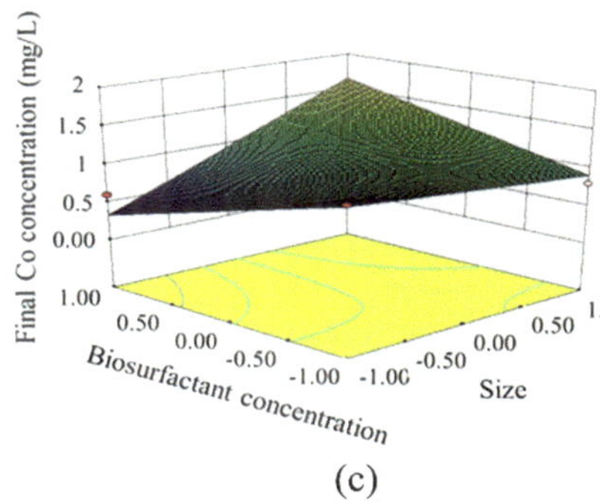

Fig. 10.10 (continued)

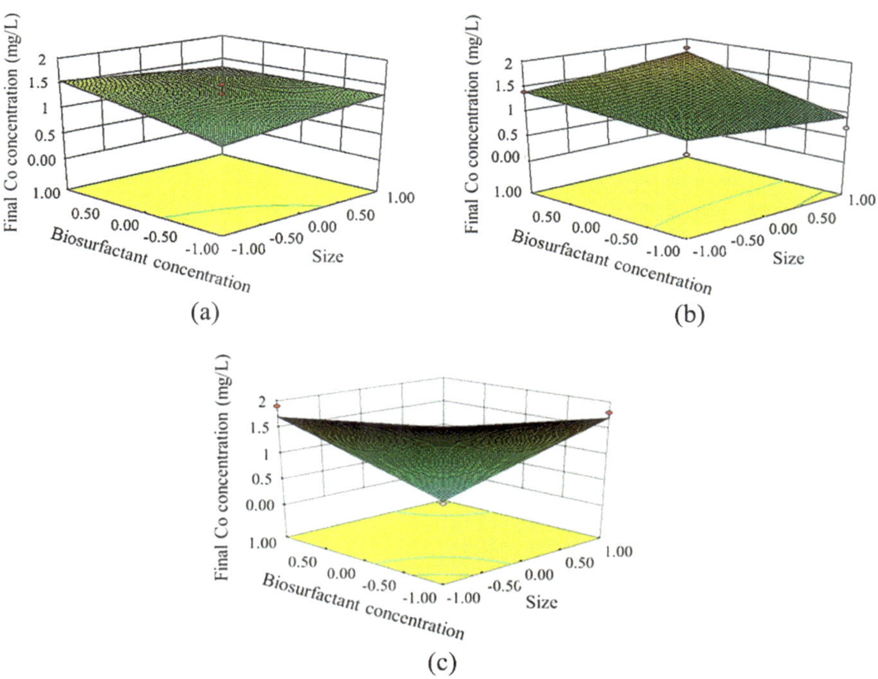

Fig. 10.11 3D Surface for interaction effects for remaining Co at **a** center point (200 mg/L), **b** low level (100 mg/L) and **c** high level (300 mg/L) of initial Co concentration with trehalose lipids

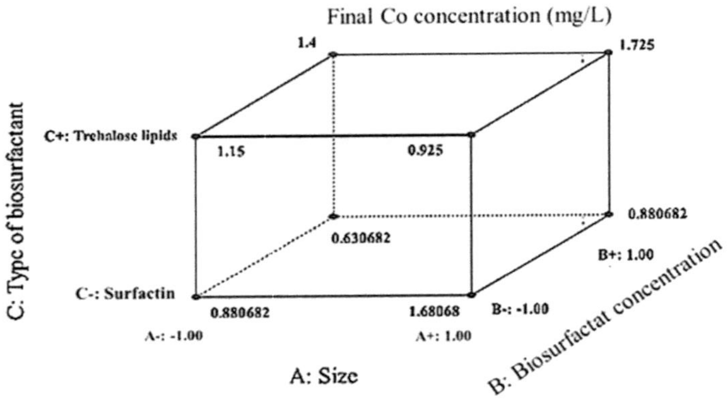

Fig. 10.12 Cube graph of actual yield applying at low concentration of initial Co

10.4 Conclusion

Saline soil is studied for the Co adsorption from aqueous solution in the presence of the biosurfactants surfactin and trehalose lipids. The effects of soil particle size, type and concentration of biosurfactant and initial Co concentration on the adsorption were studied with 22 experiments using a 2^4 full factorial design. The p-value of model (0.0008) and lack of fit (0.4372) show the model is significant with good fitting. Size of the soil particles has a positive effect on the remaining Co when surfactin and trehalose lipids are applied, larger particles result in more Co remaining in solution. However, there is an exception with large soil particles when high concentration of trehalose lipids is applied, which can be explained by changes in physicochemical properties of different-sized soil particles. When the effects of two different biosurfactants on the remaining Co are compared, more remaining Co is detected in the matrix when using trehalose lipids. High concentration of biosurfactants (>CMC), have negative and positive impact on the remaining Co concentration in the same matrix for surfactin and trehalose lipids, respectively. Increased size of the soil particles significantly increases the remaining Co in the solution when the applied biosurfactant is surfactin compared to trehalose lipids applied in the same matrix.

Statistical analysis results verify the significance of the model developed from experimental data to optimize the parameters. The optimal condition has been obtained with the size of particle of $\leq$2 mm, a 1.5 CMC of trehalose lipids and the initial Co concentration of 100 mg/L. This leads to a remaining Co of 1.73 mg/L and the desirability of 0.88. The research findings help to orient parameter interactions as they affect Co adsorption in saline soil and to predict a suitable a remediation method for Co contaminated soil.

Acknowledgements The authors would like to express their appreciation to the Canada Foundation for Innovation (CFI) and the Natural Sciences and Engineering Research Council of Canada (NSERC) for their support.

References

1. J. Acosta, B. Jansen, K. Kalbitz, A. Faz, S. Martínez-Martínez, Salinity increases mobility of heavy metals in soils. Chemosphere **85**, 1318–1324 (2011)
2. C. Adam, J. Garnier-Laplace, S. Roussel-Debet, Cobalt-60 and the Environment (2010). https://www.irsn.fr/EN/Research/publications-documentation/radionuclides-sheets/environment/Pages/Cobalt-60-environment.aspx. Accessed 3 Jan 2022
3. M.J. Anderson, P.J. Whitcomb, *RSM Simplified: Optimizing Processes Using Response Surface Methods for Design of Experiments* (Productivity Press, New York, NY, 2016)
4. N.A. Anjum, H.P. Singh, M.I.R. Khan, A. Masood, T.S. Per, A. Negi, D.R. Batish, N.A. Khan, A.C. Duarte, E. Pereira, I. Ahmad, Too much is bad—an appraisal of phytotoxicity of elevated plant-beneficial heavy metal ions. Environ. Sci. Pollut. Res. **22**, 3361–3382 (2015)
5. Y. Aşçı, M. Nurbaş, Y.S. Açıkel, Sorption of Cd(II) onto kaolin as a soil component and desorption of Cd(II) from kaolin using rhamnolipid biosurfactant. J. Hazard. Mater. **139**, 50–56 (2007)
6. Y. Aşçı, M. Nurbaş, Y.S. Açıkel, A comparative study for the sorption of Cd(II) by K-feldspar and sepiolite as soil components, and the recovery of Cd(II) using rhamnolipid biosurfactant. J. Environ. Manage. **88**, 383–392 (2008)
7. Y. Aşçı, M. Nurbaş, Y.S. Açıkel, A comparative study for the sorption of Cd(II) by soils with different clay contents and mineralogy and the recovery of Cd(II) using rhamnolipid biosurfactant. J. Hazard. Mater. **154**, 663–673 (2008)
8. Y. Aşçı, M. Nurbaş, Y.S. Açıkel, Removal of zinc ions from a soil component Na-feldspar by a rhamnolipid biosurfactant. Desalination **223**, 361–365 (2008)
9. Y. Aşçı, M. Nurbaş, Y.S. Açıkel, Investigation of sorption/desorption equilibria of heavy metal ions on/from quartz using rhamnolipid biosurfactant. J. Environ. Manage. **91**, 724–731 (2010)
10. Q. Cai, Biosurfactant production and applications in oil contaminate control. PhD thesis, Memorial University of Newfoundland, St. John's, NL, Canada (2019)
11. Q. Cai, B. Zhang, B. Chen, T. Cao, Z. Lv, Biosurfactant produced by a Rhodococcus erythropolis mutant as an oil spill response agent. Water Qual. Res. J. **51**, 97–105 (2016)
12. Q. Cai, B. Zhang, B. Chen, Z. Zhu, W. Lin, T. Cao, Screening of biosurfactant producers from petroleum hydrocarbon contaminated sources in cold marine environments. Mar. Pollut. Bull. **86**, 402–410 (2014)
13. Q. Cai, Z. Zhu, B. Chen, B. Zhang, Oil-in-water emulsion breaking marine bacteria for demulsifying oily wastewater. Water Res. **149**, 292–301 (2019)
14. J.T. Champion, J.C. Gilkey, H. Lamparski, J. Retterer, R.M. Miller, Electron microscopy of rhamnolipid (biosurfactant) morphology: effects of pH, cadmium, and octadecane. J. Colloid Interface Sci. **170**, 569–574 (1995)
15. W.-J. Chen, L.-C. Hsiao, K.K.-Y. Chen, Metal desorption from copper(II)/nickel(II)-spiked kaolin as a soil component using plant-derived saponin biosurfactant. Process Biochem. **43**, 488–498 (2008)
16. P.J. Coughtrey, D. Jackson, M. Thorne, *Radionuclide Distribution and Transport in Terrestrial and Aquatic Ecosystems: A Critical Review of Data* (AA Balkema, Netherlands, 1983)
17. E.F. Covelo, F.A. Vega, M.L. Andrade, Heavy metal sorption and desorption capacity of soils containing endogenous contaminants. J. Hazard. Mater. **143**, 419–430 (2007)
18. A.R. Dehghani-Sanij, M.R. Khani, B. Zhang, S. Narimannejad, M. Mohammadni, Water quality analysis of underground reservoirs in hot and arid regions. J. Appl. Environ. Biol. Sci **6**, 149–161 (2016)
19. A.R. Dehghani-Sanij, M.R. Khani, A. Jalali, M. Khani, S. Narimannejad, Evaluation of water quality of cisterns. Int. J. Environ. Resour. **4**, 1–8 (2015)
20. M.L. Ferreira, A.C. Casabuono, S.T. Stacchiotti, A.S. Couto, S.A. Ramirez, D.L. Vullo, Chemical characterization of Pseudomonas veronii 2E soluble exopolymer as Cd(II) ligand for the biotreatment of electroplating wastes. Int. Biodeterior. Biodegradation **119**, 605–613 (2017)

21. U. Filipkowska, M. Kuczajowska-Zadrożna, Investigation of the adsorption/desorption equilibria of Cd(II), Zn(II) and Cu(II) ions on/from immobilized digested sludge using biosurfactants. Environ. Earth Sci. **75**, 814 (2016)
22. L. Frazer, Lipid lather removes metals. Environ. Health Perspect. **108**, A320–A323 (2000)
23. C. Gong, L. Ma, H. Cheng, Y. Liu, D. Xu, B. Li, F. Liu, Y. Ren, Z. Liu, C. Zhao, K. Yang, H. Nie, C. Lang, Characterization of the particle size fraction associated heavy metals in tropical arable soils from Hainan Island, China. J. Geochem. Explor. **139**, 109–114 (2014)
24. E. Hamilton, The geobiochemistry of cobalt. Sci. Total Environ. **150**, 7–39 (1994)
25. V. Hatje, T. Payne, D. Hill, G. McOrist, G. Birch, R. Szymczak, Kinetics of trace element uptake and release by particles in estuarine waters: effects of pH, salinity, and particle loading. Environ. Int. **29**, 619–629 (2003)
26. E. Heiderscheidt, T. Leiviskä, B. Kløve, Chemical treatment response to variations in non-point pollution water quality: results of a factorial design experiment. J. Environ. Manage. **150**, 164–172 (2015)
27. K.-J. Hong, S. Tokunaga, T. Kajiuchi, Evaluation of remediation process with plant-derived biosurfactant for recovery of heavy metals from contaminated soils. Chemosphere **49**, 379–387 (2002)
28. B. Huang, Z. Li, J. Huang, L. Guo, X. Nie, Y. Wang, Y. Zhang, G. Zeng, Adsorption characteristics of Cu and Zn onto various size fractions of aggregates from red paddy soil. J. Hazard. Mater. **264**, 176–183 (2014)
29. M. Jalali, M. Majeri, Cobalt sorption–desorption behavior of calcareous soils from some Iranian soils. Chem. Erde-Geochem. **76**, 95–102 (2016)
30. A. Khataee, A. Naseri, M. Zarei, M. Safarpour, L. Moradkhannejhad, Chemometrics approach for determination and optimization of simultaneous photooxidative decolourization of a mixture of three textile dyes. Environ. Technol. **33**, 2305–2317 (2012)
31. A. Khataee, M. Zarei, R. Ordikhani-Seyedlar, Heterogeneous photocatalysis of a dye solution using supported TiO_2 nanoparticles combined with homogeneous photoelectrochemical process: molecular degradation products. J. Mol. Catal. A: Chem. **338**, 84–91 (2011)
32. A.I. Khuri, S. Mukhopadhyay, Response surface methodology. Wiley Interdiscip. Rev. Comput. Stat. **2**, 128–149 (2010)
33. K.-H. Kim, S.A. Jahan, E. Kabir, A review on human health perspective of air pollution with respect to allergies and asthma. Environ. Int. **59**, 41–52 (2013)
34. B. Kiran, K. Thanasekaran, Metal tolerance of an indigenous cyanobacterial strain, Lyngbya putealis. Int. Biodeterior. Biodegradation **65**, 1128–1132 (2011)
35. R. Kumar, D. Bhatia, R. Singh, S. Rani, N.R. Bishnoi, Sorption of heavy metals from electroplating effluent using immobilized biomass Trichoderma viride in a continuous packed-bed column. Int. Biodeterior. Biodegradation **65**, 1133–1139 (2011)
36. A. Lesniewicz, L. Gackiewcz, W. Zyrnicki, Biodegradation of metallic surgical implants investigated using an ultrasound-assisted process combined with ICP-OES and XRD. Int. Biodeterior. Biodegradation **64**, 81–85 (2010)
37. C.-H. Lin, C.-H. Lai, Y.-P. Peng, P.-C. Wu, K.-Y. Chuang, T.-Y. Yen, Y.-K. Xiang, Comparative health risk of inhaled exposure to organic solvents, toxic metals, and hexavalent chromium from the use of spray paints in Taiwan. Environ. Sci. Pollut. Res. **26**(33), 33906–33916 (2019)
38. W. Lin, L. Jing, Z. Zhu, Q. Cai, B. Zhang, Removal of heavy metals from mining wastewater by micellar-enhanced ultrafiltration (MEUF): experimental investigation and Monte Carlo-based artificial neural network modeling. Water Air Soil Pollut. **228**, 206 (2017)
39. A. Linna, P. Oksa, P. Palmroos, P. Roto, P. Laippala, J. Uitti, Respiratory health of cobalt production workers. Am. J. Ind. Med. **44**, 124–132 (2003)
40. G. Liu, J. Wang, W. Xue, J. Zhao, J. Wang, X. Liu, Effect of the size of variable charge soil particles on cadmium accumulation and adsorption. J. Soils Sediments **17**, 2810–2821 (2017)
41. L.M. Lye, Applications of DOE in Engineering and Science: A Collection of 2^6 Case Studies (2019). https://cdn.statease.com/media/public/documents/Applications_of_DOE_in_Engineering_and_Science_2019_Revised-LYE_Sept_17.pdf. Accessed 3 Jan 2021

42. D.C. Montgomery, R.H. Myers, W.H. Carter Jr., G.G. Vining, The hierarchy principle in designed industrial experiments. Qual. Reliab. Eng. Int. **21**, 197–201 (2005)
43. C.N. Mulligan, Environmental applications for biosurfactants. Environ. Pollut. **133**, 183–198 (2005)
44. C.N. Mulligan, Recent advances in the environmental applications of biosurfactants. Curr. Opin. Colloid Interface Sci. **14**, 372–378 (2009)
45. C.N. Mulligan, R.N. Yong, B.F. Gibbs, Heavy metal removal from sediments by biosurfactants. J. Hazard. Mater. **85**, 111–125 (2001)
46. R.H. Myers, D.C. Montgomery, C.M. Anderson-Cook, *Response Surface Methodology: Process and Product Optimization Using Designed Experiments* (Wiley, 2016)
47. S. Narimannejad, Review of the book 'Cisterns: sustainable development, architecture and energy.' Int. J. Ambient Energy **41**(14), 1681–1682 (2018)
48. S. Narimannejad, Adsorption Behaviours of Cobalt onto Saline Soil with/without Biosurfactants. Masters thesis, Memorial University of Newfoundland, St. John's, NL, Canada (2019)
49. S. Narimannejad, B. Zhang, L. Lye, Adsorption behavior of Cobalt onto saline soil with/without a biosurfactant: kinetic and isotherm studies. Water, Air, Soil Pollut. **230**(47) (2019)
50. V.M. Ngole-Jeme, Fire-induced changes in soil and implications on soil sorption capacity and remediation methods. Appl. Sci. **9**, 3447 (2019)
51. F.J. Ochoa-Loza, J.F. Artiola, R.M. Maier, Stability constants for the complexation of various metals with a rhamnolipid biosurfactant. J. Environ. Qual. **30**, 479–485 (2001)
52. OSHA, Air Contaminants. 29 CFR 1910.1000 [82 FR 2735, January 9, 2017] (2017). https://www.osha.gov/annotated-pels/table-z-1#notes. Accessed on 23 Aug 2023
53. C. Papelis, *Evaluation of Cobalt Mobility in Soils from the Nevada Test Site* (Nevada, University and Community College System of Nevada, Water Resources Center, Las Vegas, 1996)
54. V. Phogat, V. Tomar, R. Dahiya, Soil physical properties, in *Soil Science: An Introduction.* ed. by R.K. Rattan, J.C. Katyal, B.S. Dwivedi, A.K. Sarkar, T. Bhattachatyya, J.C. Tarafdar, S.S. Kukal (Indian Society of Soil Sciences, Delhi, India, 2015), pp.135–171
55. L. Polechońska, A. Samecka-Cymerman, Cobalt and nickel content in Hydrocharis morsus-ranae and their bioremoval from single-and binary solutions. Environ. Sci. Pollut. Res. **25**(32), 32044–32052 (2018)
56. G.F. Reed, F. Lynn, B.D. Meade, Use of coefficient of variation in assessing variability of quantitative assays. Clin. Diagn. Lab. Immunol. **9**, 1235–1239 (2003)
57. A. Regti, M.R. Laamari, S.-E. Stiriba, M. El Haddad, Use of response factorial design for process optimization of basic dye adsorption onto activated carbon derived from Persea species. Microchem. J. **130**, 129–136 (2017)
58. P. Rengasamy, World salinization with emphasis on Australia. J. Exp. Bot. **57**, 1017–1023 (2006)
59. D. Rocak, M. Kosec, A. Degen, Ceramic suspension optimization using factorial design of experiments. J. Eur. Ceram. Soc. **22**, 391–395 (2002)
60. E.Z. Ron, E. Rosenberg, Natural roles of biosurfactants. Environ. Microbiol. **3**, 229–236 (2001)
61. G. Sayyad, M. Afyuni, S.-F. Mousavi, K.C. Abbaspour, B.K. Richards, R. Schulin, Transport of Cd, Cu, Pb and Zn in a calcareous soil under wheat and safflower cultivation—a column study. Geoderma **154**, 311–320 (2010)
62. S. Shekhar, A. Sundaramanickam, T. Balasubramanian, Biosurfactant producing microbes and their potential applications: a review. Crit. Rev. Environ. Sci. Technol. **45**, 1522–1554 (2015)
63. A.K. Singh, S.S. Cameotra, Efficiency of lipopeptide biosurfactants in removal of petroleum hydrocarbons and heavy metals from contaminated soil. Environ. Sci. Pollut. Res. **20**, 7367–7376 (2013)
64. R.D.C. Soltani, A. Rezaee, H. Godini, A. Khataee, A. Hasanbeiki, Photoelectrochemical treatment of ammonium using seawater as a natural supporting electrolyte. Chem. Ecol. **29**, 72–85 (2013)
65. USEPA, *Ecological Soil Screening Levels for Cobalt* (US EPA, Washington, DC, 2005)

66. S. Wang, C.N. Mulligan, Rhamnolipid biosurfactant-enhanced soil flushing for the removal of arsenic and heavy metals from mine tailings. Process Biochem. **44**, 296–301 (2009)
67. WHO, IARC Monographs on the Evaluation of Carcinogenic Risks to Humans, Vol. 90 Human Zapillomaviruses, World Health Organization International Agency for Research on Cancer, Lyon, France (2007). https://monographs.iarc.fr/wp-content/uploads/2018/06/mono90.pdf. Accessed 25 Aug 2023
68. X. Yin, B. Gao, L.Q. Ma, U.K. Saha, H. Sun, G. Wang, Colloid-facilitated Pb transport in two shooting-range soils in Florida. J. Hazard. Mater. **177**, 620–625 (2010)
69. H. Zhang, Y. Luo, T. Makino, L. Wu, M. Nanzyo, The heavy metal partition in size-fractions of the fine particles in agricultural soils contaminated by waste water and smelter dust. J. Hazard. Mater. **248**, 303–312 (2013)
70. Z. Zhu, B. Zhang, B. Chen, Q. Cai, W. Lin, Biosurfactant production by marine-originated bacteria bacillus Subtilis and its application for crude oil removal. Water Air Soil Pollut. **227**, 328 (2016)

Chapter 11
Real-Time Optimization of Yaw Angle and Tip-Speed Ratio for a Six-Turbine Plant of NREL 5-MW Wind Turbine

Amir Hosseini, Daniel Trevor Cannon, and Ahmad Vasel-Be-Hagh

Abstract Energy loss in wind farms primarily occurs due to the aerodynamic interactions among wind turbines. Every wind turbine produces long aerodynamic wakes characterized by significant wind speed deficits and heightened turbulent kinetic energy. These wakes have a notable impact on the energy production of downstream turbines that overlap with them. Designing a wind farm that remains unaffected by the aforementioned energy losses is not feasible due to the constant variation in wind direction and short distances between the wind turbines. Using active control techniques, such as yaw control to redirect the wake or tip-speed ratio (TSR) control to weaken it, has emerged as a potential strategy to mitigate this issue partially. These techniques can adapt to changes in wind direction and speed, offering a means to address the problem to some extent. This study examines the effectiveness of implementing simultaneous yaw and TSR control in a wind farm comprising six turbines under a constant wind speed condition. Comparatively, TSR control proved to be more effective than yaw control. However, it is worth noting that the simultaneous application of both techniques resulted in even more significant energy savings. It is important to emphasize that the results are specific to the particular wind farm layout considered, and they are expected to vary if there are any changes in the farm's configuration.

Keywords Wind energy · Renewable energy · Yaw control · Tip-speed ratio · Optimization

11.1 Introduction

Based on data from the U.S. Energy Information Administration in 2022, wind turbines accounted for approximately 10.2% of the total electricity generation in the United States from utility-scale sources [1]. As outlined in the latest Wind Vision established by the U.S. Department of Energy, the target is for wind energy to con-

---

A. Hosseini · D. T. Cannon (✉) · A. Vasel-Be-Hagh
Mechanical Engineering Department, Tennessee Tech University, Cookeville 38505, TN, USA
e-mail: dtcannon42@tntech.edu

D. S-K. Ting and A. Vasel-Be-Hagh (eds.), *Engineering to Adapt*, Springer Proceedings in Energy, https://doi.org/10.1007/978-3-031-47237-4_11

tribute 20% and 35% of the nation's electricity supply by 2030 and 2050, respectively [2]. Achieving such bold goals requires more efficient wind farms, demanding strategies to address the wake loss issue as the primary source of power loss in commercial wind turbines [3].

Employing an effective, sophisticated layout optimization is crucial to minimizing the exposure of downwind turbines to the wake of their upstream counterparts [4–6]; however, there is still room for improvement by applying adaptive control strategies during operation [7]. The concept behind employing active control techniques in wind farm applications is to allow a subset of turbines to operate at a sub-optimal condition. This approach aims to preserve more power and minimize turbulence in the wind that reaches the remaining turbines. Identifying the specific subset of turbines and their deviation from optimal conditions requires an optimization process. This process ensures that the energy loss experienced by these turbines is outweighed by the gain achieved by their downstream counterparts, ultimately leading to an overall improvement in the wind farm's production.

One of the active control strategies for redirecting a wind turbine wake away from the downstream turbines is yaw control [8, 9]. Yaw control entails deliberately inducing a yaw misalignment in a turbine, reducing power production for that specific turbine (upstream loss). However, this strategy serves to deflect the wake, thereby diminishing the exposure of downstream turbines to upstream wakes and augmenting their power production (downstream gain). When implemented accurately, the downstream gains can surpass the upstream losses, leading to an overall increase in power production.

An additional effective strategy for wake control is the manipulation of the tip-speed ratio (TSR), as introduced by Hosseini et al. [10]. This approach focuses on reducing the wind speed deficit in upstream wakes, thereby increasing the amount of available power within the wake for downstream turbines to capture. A wind turbine's efficiency depends on the TSR. At a specific TSR, the turbine attains its maximum efficiency. Presently, the industry maintains all turbines at this optimal TSR, assuming that maximizing each turbine's individual performance will result in the overall efficiency of the wind farm being maximized as well. However, this assumption is flawed, as demonstrated by Hosseini et al. [10], who illustrated that adjusting the TSR for each turbine in response to wind speed and direction changes can result in a greater overall output for the wind farm.

This study investigates the efficacy of concurrently employing both yaw and TSR control strategies. The six-turbine farm, originally adapted by Gebraad et al. [11], serves as the basis for the investigation. The wind direction ranges from 0° to 360° in 5° increments, while the wind speed remains constant at 8 m/s, consistent with the conditions employed by Gebraad et al. [11]. Section 11.2 describes the study methodology, while the results and their analysis are presented in Sect. 11.3.

11.2 Methodology

11.2.1 Wake Modeling

The FLORIS (FLOw Redirection and Induction in Steady State) [12], a three-zone wake model, was employed in the present study due to its ability to account for both wake deflection and expansion. By incorporating the aerodynamic characteristics of turbines, the FLORIS model accurately calculates the wind speed deficit in the downstream region. The FLORIS model is particularly suitable for real-time yaw optimization since it can effectively account for wake deflection resulting from yaw misalignment. Suppose power and thrust coefficients under ideal non-yawed conditions where the turbine's rotor is normal to the incoming wind are $c_{p,non-yawed}$ and $c_{t,non-yawed}$. One can use FLORIS to calculate the turbine's actual power coefficient when subjected to a yaw misalignment of γ and the wake deflection caused by this yaw misalignment denoted as δy_{wake}, as

$$c_p(\gamma) = F(c_{p,non-yawed}, \gamma) \tag{11.1}$$

$$\delta y_{wake} = G(c_{t,non-yawed}, \gamma, x) \tag{11.2}$$

where x is the distance from the rotor downstream in the wind direction, and F and G are the functions defined in [11]. In their study, Gebraad et al. [11] computed $c_{p,non-yawed}$ and $c_{t,non-yawed}$ by assuming a constant induction factor of $a = 1/3$ and using the formulas $c_p = 4a(1-a)^2$ and $c_t = 4a(1-a)$. However, in our case, as we aimed to implement a combined yaw/TSR (Tip Speed Ratio) optimization, we opted to utilize the manufacturer-provided c_p and c_t curves (see Fig. 11.1). This approach allows for a direct relationship between TSR and the non-yawed power and thrust coefficients.

11.2.2 Optimization

This study utilized a particle swarm optimization technique, which closely resembled the methodology described by Hosseini et al. [10]. To avoid redundancy, we direct readers to refer to their work [10] for detailed information on the methodology.

11.3 Results and Discussions

Figure 11.2 depicts the studied wind farm. This wind farm is identical to that considered by Gebraad et al. [11], with turbine labels resembling those of Gebraad et al. This layout provides several directions of alignment through which wind direction

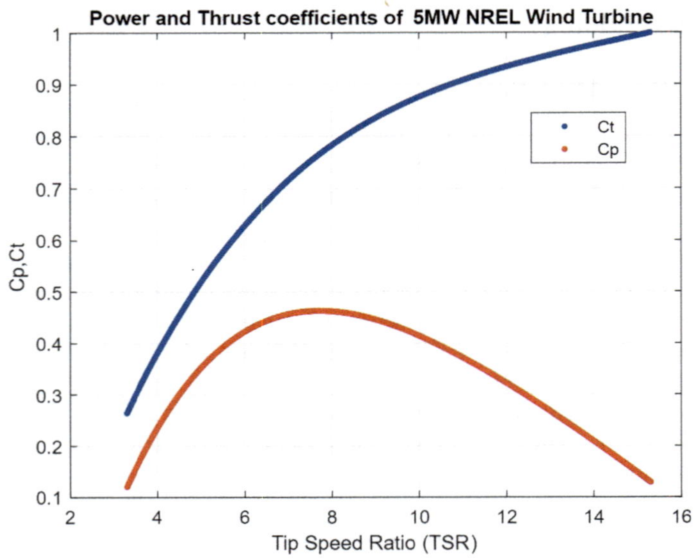

Fig. 11.1 NREL 5 MW power and thrust coefficient curves

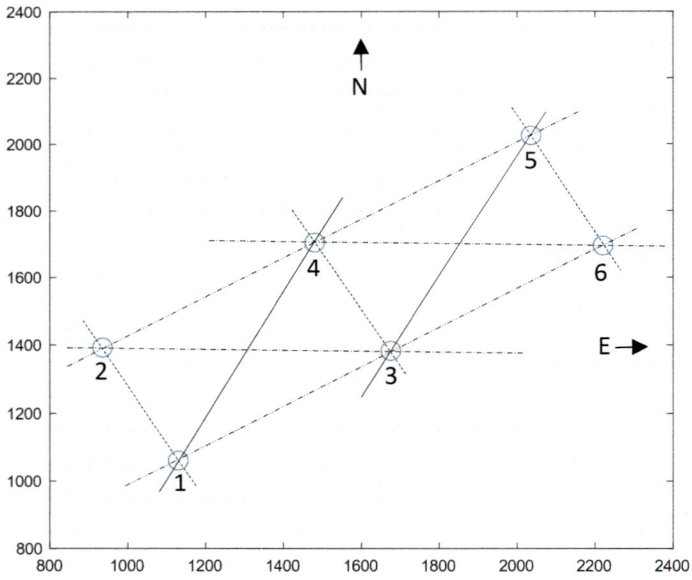

Fig. 11.2 The layout and the alignment directions shown for wind farm rotation of 0°

Table 11.1 Validating the MATLAB implementation of FLORIS using a PSO algorithm at three specific wind directions (=three farm rotations). For plant rotation of 0° the wind direction is southwest, aligned with the column formed by turbines 1, 3, and 6

		Optimal Yaw Angles						
		Turbine 1	Turbine 2	Turbine 3	Turbine 4	Turbine 5	Turbine 6	Change in Power (MW)
The plant rotated 0 degrees	MATLAB (present work)	27.4	25.9	40	40	0	0	6.28 → 7.76 **(+23.6%)**
	Original (Gebraad et al.)	25.8	25.1	39.8	39.7	0.5	0.4	6.34 → 7.66 **(+20.8%)**
The plant rotated 5 degrees	MATLAB (present work)	18.3	17.8	23.7	23.1	0	0	9.59 → 11.17 **(+16.5%)**
	Original (Gebraad et al.)	19.0	19.0	39.8	39.7	0.45	0.35	8.75 → 9.99 **(14.2%)**
The plant rotated 10 degrees	MATLAB (present work)	6.6	6.7	6.8	6.0	0	0	13.61 → 13.89 **(+2.0%)**
	Original (Gebraad et al.)	6.2	3.4	4.7	6.3	0	0	11.04 → 11.22 **(1.6%)**

aligns with a column of two or three turbines. These directions matter most since they lead to maximum wake interactions. In other words, alignment directions cause maximum overlap between the upstream wake and downstream rotor areas; hence, the adaptive control strategies are expected to be most effective in these directions.

First, we optimized the farm's yaw angle alone and compared the results versus those presented by Gebraad et al. [11]. This comparison was essential to verify the accurate implementation of the FLORIS model. In our work, FLORIS was implemented in MATLAB to facilitate its integration with the built-in Particle Swarm Optimization (PSO) algorithm. This approach eliminated the necessity of programming PSO separately. It is important to note that Gebraad et al. [11] did not utilize PSO in their study. Therefore, we anticipated certain variations between our results and theirs due to this difference in methodology. Results are presented in Table 11.1. This algorithm optimized the farm's yaw angles in every wind direction (5°) at 8 m/s. As expected, the algorithm exclusively applied yaw misalignment to the upstream turbines, as altering the yaw angle of a downstream turbine would not provide any advantages to the other turbines and would only result in losses for the yawed turbine without benefiting the overall system. The optimization results are illustrated in Fig. 11.3, displaying the optimal yaw angles for respective wind directions. It is important to note that this figure presents only four wind directions, while the analysis encompassed 72 wind directions.

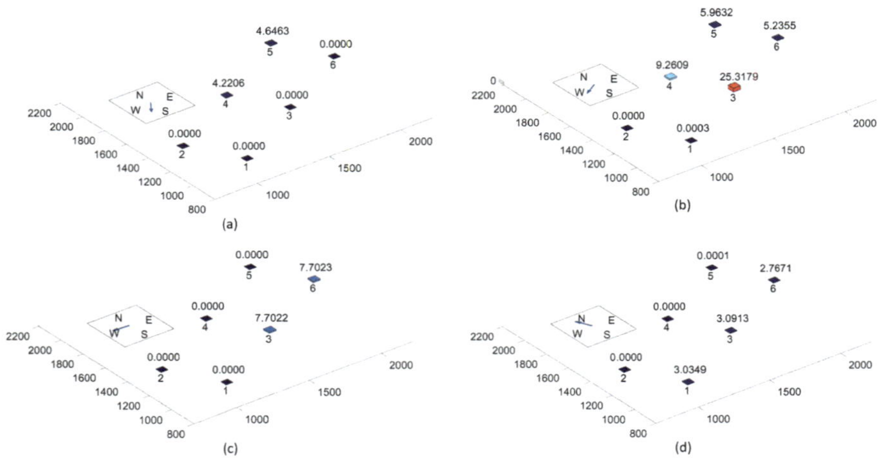

Fig. 11.3 Yaw optimization: optimal yaw angles in sample directions of alignment

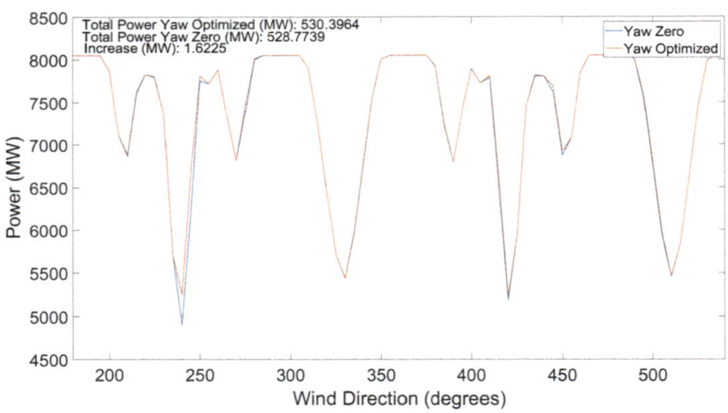

Fig. 11.4 Comparing the power production: original farm versus the farm with yaw optimization

As evident from the results, the algorithm recommends applying yaw misalignment exclusively to turbines that have a corresponding turbine located in their wake. In contrast, the last row of turbines remains unaffected by yaw adjustments since redirecting their wake would not provide any benefits to other turbines. Figure 11.4 depicts the power production variations across different wind directions during the implementation of real-time yaw optimization. Note that the wind direction is with respect to the meteorological convention, with the north being at 0°. The analysis demonstrated a total power production increase of 1.62 MW from the yaw optimization.

We exclusively optimized the farm's TSR before applying the combined yaw/TSR optimization. Similar to the yaw optimization, a change in TSR was recommended

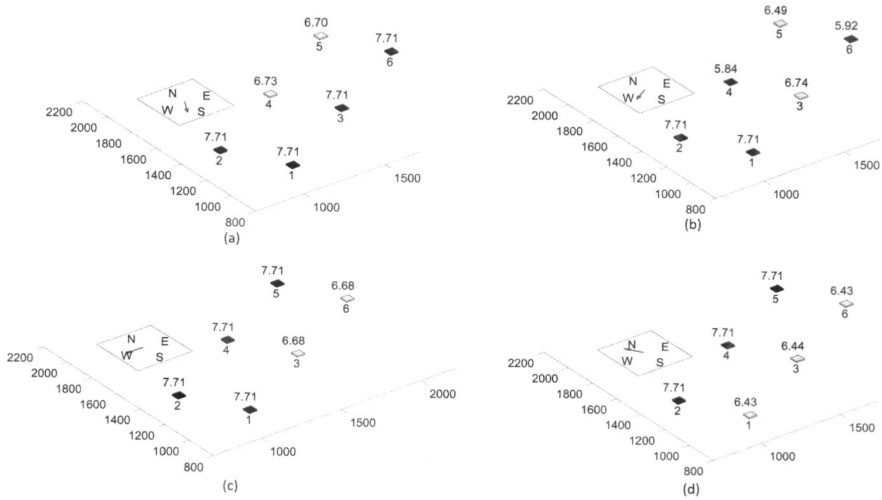

Fig. 11.5 TSR optimization: optimal TSRs in sample directions of alignment

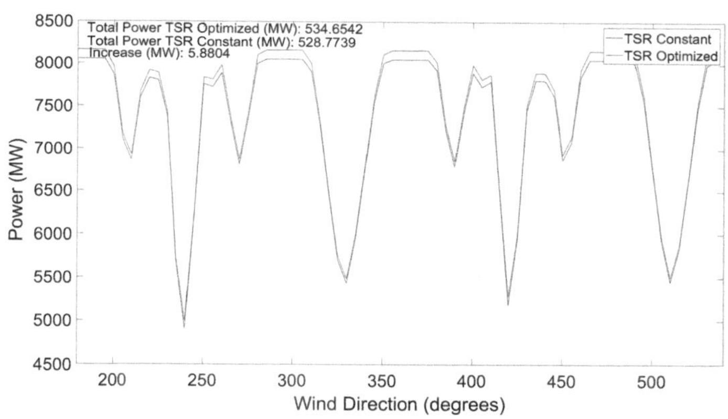

Fig. 11.6 Comparing the power production: original farm versus the farm with TSR optimization

only for turbines whose wake impacted one or two downstream turbines. The algorithm determined a reduction in the TSR of these turbines, thereby diminishing their efficiency, to facilitate the downstream turbines in capturing more power from the upstream wake. Figure 11.5 illustrates the optimal TSR values corresponding to the same four wind directions depicted in Fig. 11.3. The original farm maintains a TSR of 7.71 across all wind directions. In contrast, the optimized farm lowers the TSR of turbines that have other turbines in their wake by up to 25%. Consider Fig. 11.5a, for instance. The algorithm dropped TSR at turbines 4 and 5 because their wake affects turbines 1 and 3, respectively. Figure 11.6 showcases the power production for both the original farm and the farm implementing real-time TSR control. As a result of

TSR optimization, the overall power production witnessed a substantial increase of 5.9 MW, surpassing the impact of yaw optimization. TSR control offers additional benefits beyond improving a wind farm's energy performance. As highlighted in [10], implementing TSR control resulted in a lower average TSR for the farm, subsequently reducing the overall rotational speed of the turbines. This decrease in blade speed contributes to mitigating concerns related to bird collisions and reduces noise generation and erosion and extends the turbines life time.

In the final phase, we performed a combined TSR/yaw optimization. As anticipated, the optimal values obtained for yaw and TSR differed from those achieved when either yaw or TSR was optimized independently. The total gain resulting from the combined optimization was found to be 7.65 MW, slightly exceeding the cumulative gains obtained through individual yaw and TSR optimizations (5.90 + 1.62 = 7.52 MW). Detailed results of the combined yaw/TSR optimization can be observed in Figs. 11.7 and 11.8. Although this 1.45% improvement may not carry economic significance for the small wind farm studied in this context, it holds substantial value at a utility-scale wind farm. To illustrate, a mere 1.45% boost in output at a compact utility-scale wind farm like Lillgrund, comprising 48 turbines, each with a capacity of 2.3 MW, could result in annual revenue exceeding 1.5 million dollars, assuming 23 cents per kilowatt-hour (kWh). As elaborated in our paper, we anticipate even more pronounced enhancements in utility wind farms, owing to the presence of numerous wind directions characterized by substantial wake losses, making the implementation of wake control strategies notably more effective.

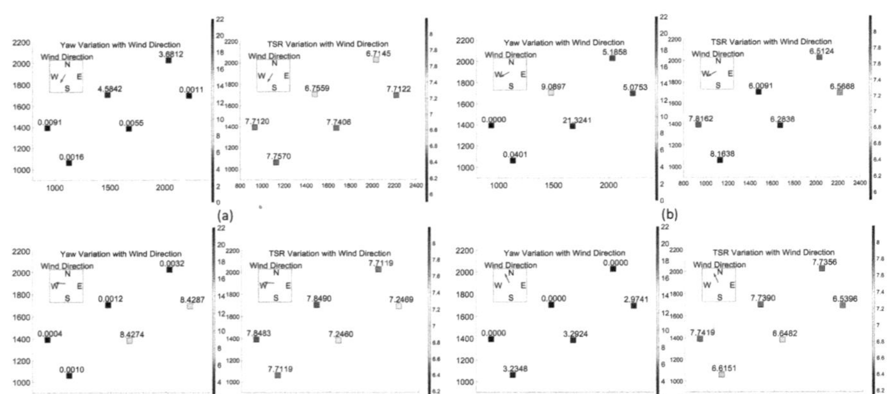

Fig. 11.7 A combined yaw/TSR optimization: optimal yaw and TSR values for sample wind directions

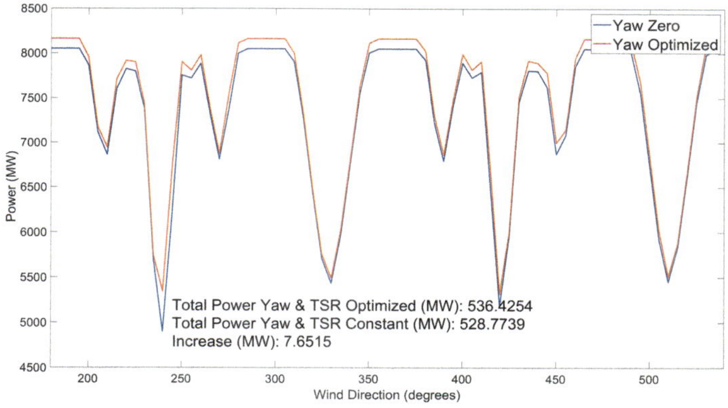

Fig. 11.8 Farms power production with and without combined yaw/TSR control

11.4 Conclusion

Real-time combined yaw/TSR optimization was performed for a six-turbine plant, encompassing all wind directions with 5° increments while maintaining a constant wind speed of 8 m/s. The findings of this study demonstrated that, when examined individually, optimizing TSR appeared more effective than optimizing yaw. However, the combined optimization proved superior, surpassing the power gains achieved by summating individual yaw and TSR optimizations. It was observed that the directions benefiting the most from yaw and TSR enhancements were identical. This aligns with intuition, as both mechanisms target the same issue and demonstrate greater effectiveness in directions where the problem is more pronounced. The results presented here would change with the farm's layout. It is worth noting that the effectiveness of the combined yaw/TSR optimization would be even more pronounced in utility-scale farms with more alignment directions compared to the relatively simple six-turbine setup investigated here. Moreover, the consideration of multiple wind speeds, rather than solely relying on a single wind speed (8 m/s) as in this study, would influence the effectiveness of the proposed strategy. By employing wind direction increments smaller than 5°, even more substantial benefits can be achieved. In future research, we intend to explore the performance of the suggested combined yaw/TSR optimization approach in a utility-scale farm. This analysis will consider 360 wind directions with 1° increments and encompass all wind speeds within the range from the cut-in to the cut-out limits. This upcoming research, to be published as a journal article in the near future, will also include a segment focusing on the economic analysis of adopting the proposed approach, aiding farm owners in their decision-making process.

References

1. United States Energy Information Administration, Wind explained: Electricity generation from wind, https://www.eia.gov/energyexplained/wind/electricity-generation-from-wind.php. Accessed: 20 April 2023
2. National Renewable Energy. Lab, *Wind vision: A new era for wind power in the united states, Technical Report DOE/GO-102015-4557* (U.S. Department of Energy, Golden, CO, United States, 2015)
3. C.L. Archer, A. Vasel-Be-Hagh, C. Yan, S. Wu, Y. Pan, J.F. Brodie, A.E. Maguire, Review and evaluation of wake loss models for wind energy applications. Appl. Energy **226**, 1187–1207 (2018)
4. M. Y. Mahfouz, P.-W. Cheng, A passively self-adjusting floating wind farm layout to increase the annual energy production. Wind Energy **26**(3), 251–265 (2023). https://doi.org/10.1002/we.2797
5. Y. Liu, The effect of vertical arrangement on performance and wake characteristics of two tandem offshore wind turbines under various operating conditions. Energy Convers. Manage. **278** (2023). https://doi.org/10.1016/j.enconman.2023.116743
6. L. Cao, M. Ge, X. Gao, B. Du, B. Li, Z. Huang, Y. Liu, Wind farm layout optimization to minimize the wake induced turbulence effect on wind turbines. Appl. Energy **323**, 8 (2022)
7. R. Nash, R. Nouri, A. Vasel-Be-Hagh, Wind turbine wake control strategies: A review and concept proposal. Energy Convers. Manage. **245**, 114581 (2021). https://doi.org/10.1016/j.enconman.2021.114581
8. M. Zhao, S. Chen, K. Wang, X. Wu, R. Zha, Effect of the yaw angle on the aerodynamics of two tandem wind turbines by considering a dual-rotor wind turbine in front. Ocean Eng. **283** (2023). https://doi.org/10.1016/j.oceaneng.2023.114974
9. C.L. Archer, A. Vasel-Be-Hagh, Wake steering via yaw control in multi-turbine wind farms: recommendations based on large-eddy simulation. Sustain. Energy Technol. Assess. **33**, 34–43 (2019). https://doi.org/10.1016/j.seta.2019.03.002
10. A. Hosseini, D.T. Cannon, A. Vasel-Be-Hagh, Tip speed ratio optimization: more energy production with reduced rotor speed. Wind **2**(4), 691–710 (2022). https://doi.org/10.3390/wind2040036.
11. P.M.O. Gebraad, F.W. Teeuwisse, J.W. van Wingerden, P.A. Fleming, S.D. Ruben, J.R. Marden, L.Y. Pao, Wind plant power optimization through yaw control using a parametric model for wake effects-a cfd simulation study. Wind Energy **19**(1), 95–114 (2016). https://doi.org/10.1002/we.1822
12. U.S. National Renewable Energy Lab, Floris, https://github.com/NREL/floris (version v3.4: May 16, 2023)